Smart Sensors, Measurement and Instrumentation

Volume 32

Series editor

Subhas Chandra Mukhopadhyay
Department of Engineering, Faculty of Science and Engineering
Macquarie University
Sydney, NSW
Australia
e-mail: subhas.mukhopadhyay@mq.edu.au

More information about this series at http://www.springer.com/series/10617

Chinthaka P. Gooneratne
Bodong Li · Max Deffenbaugh
Timothy Moellendick

Instruments, Measurement Principles and Communication Technologies for Downhole Drilling Environments

Springer

Chinthaka P. Gooneratne
Drilling Technology Division
EXPEC ARC
Dhahran, Saudi Arabia

Bodong Li
Drilling Technology Division
EXPEC ARC
Dhahran, Saudi Arabia

Max Deffenbaugh
Aramco Research Center
Houston, TX, USA

Timothy Moellendick
Drilling Technology Division
EXPEC ARC
Dhahran, Saudi Arabia

ISSN 2194-8402 ISSN 2194-8410 (electronic)
Smart Sensors, Measurement and Instrumentation
ISBN 978-3-030-04899-0 ISBN 978-3-030-04900-3 (eBook)
https://doi.org/10.1007/978-3-030-04900-3

Library of Congress Control Number: 2018962923

This Springer imprint is published by the registered company Springer Nature Switzerland AG
The registered company address is: Gewerbestrasse 11, 6330 Cham, Switzerland

Preface

This book presents a complete review of the unique instruments and the communication technologies utilized in downhole drilling environments. These instruments and communication technologies play a critical role in drilling hydrocarbon wells safely, accurately and efficiently into a target reservoir zone by acquiring information about the surrounding geological formations as well as providing directional measurements of the wellbore. Research into instruments and communication technologies for hydrocarbon drilling has not been explored by researchers to the same extent as other fields, such as biomedical, automotive and aerospace applications. Therefore, this book aims to serve as an opportunity for researchers to truly understand how instruments and communication technologies can be used in a downhole environment and to provide fertile ground for research and development in this area. A look ahead, discussing other technologies such as micro-electromechanical-systems (MEMS) and fourth industrial revolution technologies such as automation, the industrial internet of things (IIoT), artificial intelligence, and robotics that can potentially be used in the oil/gas industry are also presented, as well as requirements still need to be met in order to deploy them in the field.

Chapter 1 describes the drilling process to extract hydrocarbons from deep reservoirs and the challenges encountered due to the harsh environment downhole. Chapter 2 describes in detail the wide range of downhole parameters measured by currently available wireline (WL) and logging-while-drilling (LWD) instruments, their methods of measurement and the interpretation of logging data, which provides insight into rock formations, reservoirs and the downhole environment. Chapter 3 focuses on measurement-while-drilling (MWD) instruments for navigating a well in different three dimensional (3D) directions and obtaining drilling dynamics data for optimizing the drilling process. Chapter 4 presents the various techniques utilized by seven different logging tools, resistivity, induction, dielectric, spectral azimuthal gamma, pulsed neutron, nuclear magnetic resonance, and acoustic, to interpret downhole measurement data and provide information to the driller to optimize the drilling process. Chapter 5 presents the different telemetry techniques utilized to transmit data computed downhole to the surface, while drilling, in real-time, called 'uplink' communication and also from the surface to downhole, called 'downlink' communication. An overview of the methods employed to power instruments in inaccessible down-hole environments is also provided. Chapter 6 describes the fabrication and packaging procedures to make instruments rugged, compact, highly reliable and accurate for downhole measurements. Chapter 7 reports novel instrument technologies tested in the field recently, which pave the way for unique data acquisition and drilling methods. Finally, a discussion on potential technologies and their current status is presented in Chap. 8.

The challenges in designing sensors and instrumentation (S&I) for the harsh environments encountered downhole are unique and significant. Therefore, the authors believe the proposed book would be of interest not only to professionals, academics and students working in the realm of oil/gas but also to the general audience in the area of S&I. Postgraduates and researchers working in oil/gas companies and harsh/severe sensing environments, researchers working in commercial oil/gas companies and academic institutions that offer courses in downhole sensing and/or sensing in harsh/challenging/severe sensing environments will find the material in this book covers the whole spectrum of downhole logging tools, navigation tools, downhole communication, power supplies, fabrication and packaging, and new technologies such as MEMS/Mobile devices in downhole environments.

The chapters have also been carefully woven so that it also serves as a fundamental text for

(i) S&I researchers from other industries crossing into oil/gas, who would find this a fundamental text for understanding the downhole sensing environment.
(ii) MEMS/NEMS researchers, as research into the application of microsystems has not been explored to the same extent as other applications, such as biomedical, automotive and aerospace applications.
(iii) Researchers/Companies working in the fourth industrial revolution (4IR) technologies such as automation, artificial intelligence, the internet of things and robotics, as the transformation of the oil/gas industry to be in alignment with the 4IR has already gained traction.

The authors would like to acknowledge the Society of Petrophysicists and Well Log Analysts (SPWLA), Oilfield Review, the Society of Petroleum Engineers (SPE), Sensors Journal and the Journal of Shock and Vibration for some of the figures reproduced and published in this paper, which are originally from papers published by them. We would also like to acknowledge Dr. Paul Glover's notes in petrophysics, Directional drilling blogspot and Crain's petrophysical handbook for making a wealth of information freely available online. Last but not least, a special thank you to our families for their constant encouragement and support.

Dhahran, Saudi Arabia — Chinthaka P. Gooneratne
Dhahran, Saudi Arabia — Bodong Li
Houston, USA — Max Deffenbaugh
Dhahran, Saudi Arabia — Timothy Moellendick

Contents

Bodong Li received the B.E. degree in mechanical engineering from Jilin University, Changchun, China, and Tomsk Polytechnic University, Tomsk, Russia, in 2009, and the M.Sc./Ph.D. degree in mechanical engineering and electrical engineering respectively from the King Abdullah University of Science and Technology (KAUST), Thuwal, Saudi Arabia, in 2011 and 2014. He joined Saudi Aramco EXPEC Advanced Research Center in 2015. Currently he is working as a research engineer on the development of downhole technologies for drilling and robotic technologies for oil and gas applications.

Max Deffenbaugh is the Sensors Technology Leader at the Aramco Research Center in Houston, TX. He oversees a team of scientists and engineers who turn concepts for new oilfield instruments into field-proven prototypes. He has more than 20 years of research experience in the oil and gas industry. He worked for ExxonMobil for 16 years at their Upstream Research Center in Houston, TX and at their Corporate Strategic Research Lab in Annandale, NJ. He joined Aramco Services Company in 2013 to found and lead the Sensors Development Team. He holds a Ph.D. in Electrical Engineering and Computer Science from the Massachusetts Institute of Technology. He is a member of the Institute of Electrical and Electronics Engineers (IEEE) and the Society of Petroleum Engineers (SPE). He has authored more than 90 conference papers, peer reviewed journal articles, patents and patent applications.

Timothy Moellendick serves as Chief Technologist for the Drilling Technology Division of Saudi Aramco's EXPEC Advanced Research Center. With more than twenty five years of oil and gas industry experience, he leads a team of world class researchers in developing the next generation of drilling technology required by Saudi Aramco's Drilling and Workover stakeholders. He is considered the industry expert in Casing and Liner Drilling applications and engineering. In his previous role as Director of Technology for Schlumberger, he was responsible for growing the technical and

Contents

About the Authors

Chinthaka P. Gooneratne is the Sensors & Instrumentation (S&I) Lead and the Technical Leader for the Internet of Things in the Drilling Technology Team at EXPEC Advanced Research Center, Saudi Aramco. He has over 10 years of experience in creating, leading and managing programs designed to commercialize innovative S&I systems and developing emerging markets that utilize advanced S&I. He is a Senior Member of the Institute of Electrical and Electronics Engineers (IEEE) and has authored 10 patents/patent applications, 34 peer-reviewed journal papers, 40 conference papers and 5 book chapters. He received the B.Eng. (Hons) degree in Information and Telecommunication engineering, and the M.Eng. degree in the area of Electromagnetics (First Class Hons), from Massey University, New Zealand, in 2004 and 2005 respectively. He received the Monbukagakusho scholarship from the government of Japan in 2006 and graduated with a Ph.D. in Electrical Engineering from Kanazawa University in 2009. He was the recipient of the Kanazawa University President's Award for his work on sensor development for hyperthermia cancer treatment. He was a Postdoctoral Fellow at King Abdullah University of Science and Technology (KAUST) from 2010 to 2013, where he developed novel magnetic microsystem prototypes for biomedical applications.

Bodong Li received the B.E. degree in mechanical engineering from Jilin University, Changchun, China, and Tomsk Polytechnic University, Tomsk, Russia, in 2009, and the M.Sc./Ph.D. degree in mechanical engineering and electrical engineering respectively from the King Abdullah University of Science and Technology (KAUST), Thuwal, Saudi Arabia, in 2011 and 2014. He joined Saudi Aramco EXPEC Advanced Research Center in 2015. Currently he is working as a research engineer on the development of downhole technologies for drilling and robotic technologies for oil and gas applications.

Max Deffenbaugh is the Sensors Technology Leader at the Aramco Research Center in Houston, TX. He oversees a team of scientists and engineers who turn concepts for new oilfield instruments into field-proven prototypes. He has more than 20 years of research experience in the oil and gas industry. He worked for ExxonMobil for 16 years at their Upstream Research Center in Houston, TX and at their Corporate Strategic Research Lab in Annandale, NJ. He joined Aramco Services Company in 2013 to found and lead the Sensors Development Team. He holds a Ph.D. in Electrical Engineering and Computer Science from the Massachusetts Institute of Technology. He is a member of the Institute of Electrical and Electronics Engineers (IEEE) and the Society of Petroleum Engineers (SPE). He has authored more than 90 conference papers, peer reviewed journal articles, patents and patent applications.

Timothy Moellendick serves as Chief Technologist for the Drilling Technology Division of Saudi Aramco's EXPEC Advanced Research Center. With more than twenty five years of oil and gas industry experience, he leads a team of world class researchers in developing the next generation of drilling technology required by Saudi Aramco's Drilling and Workover stakeholders. He is considered the industry expert in Casing and Liner Drilling applications and engineering. In his previous role as Director of Technology for Schlumberger, he was responsible for growing the technical and operational knowledge base used to develop, plan and execute this technology worldwide. He has also held drilling operations and engineering positions including Senior Drilling Engineer, Drilling Manager for North America, Senior Field Engineer/Directional Driller and Operations Coordinator for the Gulf Coast of Mexico.

Chapter 1
Drilling Hydrocarbon Wells

Abstract This chapter describes the general process of drilling a hydrocarbon well and the unique and significant challenges encountered while drilling due to the harsh and challenging environment downhole. In addition to the high temperature, high pressure, highly acidic and corrosive environments encountered in deep wells, there are also the destructive, dynamic conditions, such as high torque, shock and vibration, created by the drill bit grinding and penetrating through rock formations, which makes drilling challenging. The chapter gives an insight into why logging and directional instruments are not only required to acquire measurements but also to provide an accurate representation of the well while drilling over extended periods of time, by surviving the downhole environment.

1.1 Introduction

Hydrocarbon (oil/gas) exploration and production involves drilling thousands of feet into the ground, through rock formations that have unique physical/chemical characteristics, to access hydrocarbon reservoirs that have been formed over millions of years. The first exploration and use of hydrocarbons date back to 500 BC when the Chinese extracted oil from wells over 100 ft deep [1]. However, with the European enlightenment that led to the invention of the internal combustion engine hydrocarbons became a key pillar driving the second industrial revolution, creating an impact not only on technology but also on politics, society and the world economy that has continued to this day [2]. Apart from a few downturns there has been a steady increase in hydrocarbon production since the 1900s to keep up with the increasing demand for energy, and hydrocarbons are still the critical fuel source that powers industry and transportation.

Drilling a well is a challenging procedure due to the inability to see the trajectory of the well, rock formations and the harsh environment downhole. Moreover, the drilling instruments that are lowered into a wellbore are inaccessible from the surface. Due to drillers 'drilling in the dark' well logging/directional instruments with sensors are lowered into the wellbore to provide information on the downhole

C. P. Gooneratne et al., *Instruments, Measurement Principles and Communication Technologies for Downhole Drilling Environments*, Smart Sensors, Measurement and Instrumentation 32, https://doi.org/10.1007/978-3-030-04900-3_1

environment and the drilling process, which aids the driller on the surface to map and visualize the formations and the well trajectory, and to keep track of the mechanical parameters of the drilling assembly. Downhole data obtained by these instruments play a critical role in the successful and safe drilling of a well into a target hydrocarbon reservoir. Well logs related to hydrocarbon exploration began in 1927 when Conrad and Marcel Schlumberger obtained the first experimental electric log, which had its basis on systematic studies of electrical surveying along the surface of the earth [3]. The first electric log measured the electrical potential between a probe at the surface and another probe lowered to different depths inside a well. Figure 1.1a shows a potentiometric device used for such logging and Fig. 1.1b a truck with recording equipment. Since this initial foray into obtaining

Fig. 1.1 Early logging instruments. **a** Potentiometric device used for electrical logging. **b** A truck with equipment to record the acquired data [3]

information on hydrocarbon wells, logging methods and techniques have advanced to the point where real-time data on downhole parameters can be obtained while drilling a well.

1.2 Drilling Principles

1.2.1 The Drilling Process

Prior to drilling wells prospective hydrocarbon fields are identified by seismic evaluation [4, 5]. This is followed by drilling exploration wells to gain information on the field and reservoir, such as formation types, temperature and pressure gradients, permeability, porosity, the existence, and the type of fluids and gases present, in the reservoir, along with the depth, boundaries and extent of the reservoir. Cores (sample of rock columns) are also cut and captured from the wellbore using special coring bits and retrieved at the surface [6]. Cores provide valuable information on the rocks penetrated by drilling and their pore contents. All this data is used to design a drilling program for drilling a well to a known target depth to access and, produce from, the hydrocarbon reservoir.

Figure 1.2a shows a drilling rig assembled to drill a well. During the rig-up process the derrick, a mast held by cradles, is placed in an upright position. A travelling block, which can move the drill string assembly up and down is then connected to the top of the derrick, the crown block, by a set of wire ropes. Fixed below the travelling block is a top drive with an electric motor that rotates the drilling assembly from the top, where the drilling assembly consists of hollow steel drill pipes with a drill bit at the bottom [7, 8]. The drill bit is a cutting tool that rotates and penetrates through rock formations below the surface under the combined effects of axial and longitudinal loads and the rotation of the drill string assembly [9, 10]. In some situations the rotation of the drill bit is also driven by a hydraulic motor along with the rotation from the surface. Three drill pipes connected together, 90 ft in length and called a stand, are rotated and lowered into the wellbore to penetrate into the rock formations. This process is repeated until the target well depth is reached. The general sequence of drilling a well is shown in Fig. 1.2b. While a well is being drilled fluid pumps and a circulating system ensure drilling fluid is pumped from the surface, through the drill string assembly, out the drill bit and back up the annulus, which is the space between the drill pipe and the formations, as shown in Fig. 1.2b-(i). The density of the drilling fluid must be higher than the formation pressures to prevent formation fluids from entering the annulus and flowing to the surface but lower than the mechanical strength of the formation, as a higher density may fracture the formation creating a path for the drilling fluids to go into the formation [7, 8, 11]. Apart from well control, drilling fluids also cool the bit and perform the function of lifting rock cuttings from the bottom of the hole, up the annulus and to the surface to be filtered out and treated before it is pumped down the drill string again, as shown in Fig. 1.2a.

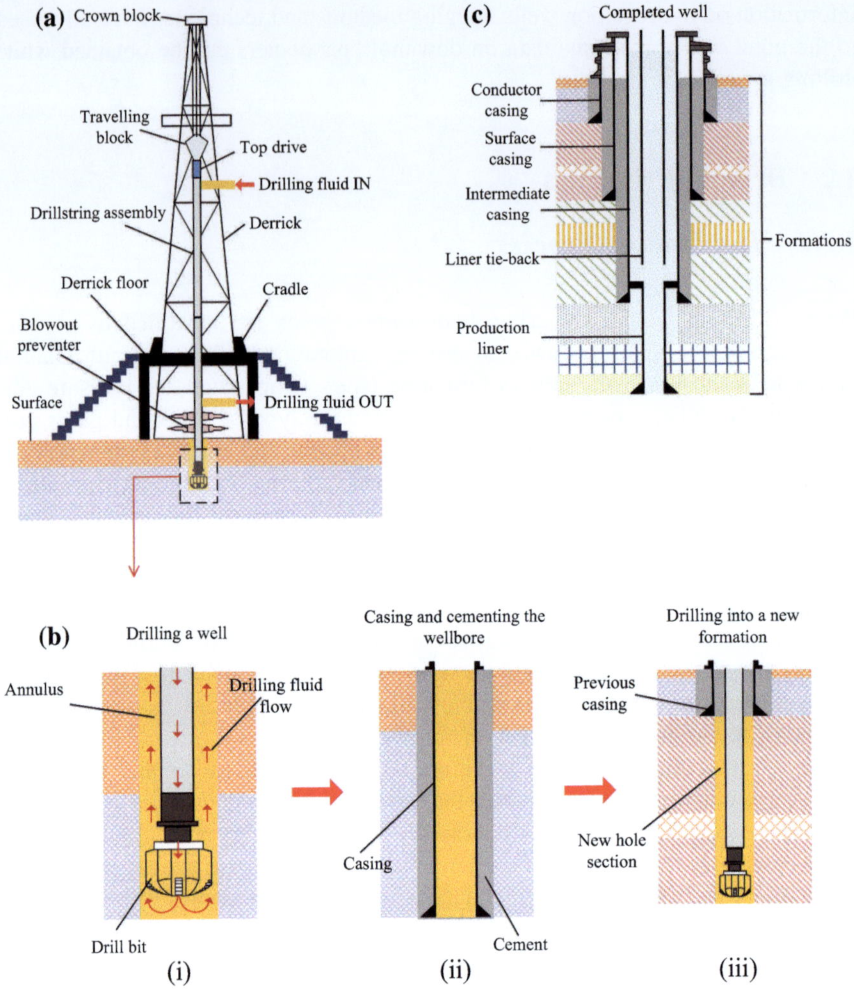

Fig. 1.2 **a** A fully assembled drilling rig with a drilling assembly and a drill bit penetrating rock formations below the surface. **b** The drilling process. (i) A drill bit penetrates formations and drills a section. (ii) A steel casing is lowered and cemented to isolate the well from the formations. (iii) The drill bit continues to drill the next section. **c** A completed well with several cemented casings ready for production

Once the well is drilled to the target depth, the drilling assembly is pulled out of the well and a steel casing is lowered and lined into the wellbore to support the sides of the wellbore, as shown in Fig. 1.2b-(ii). Once the casing is lowered to the bottom, a cement slurry is pumped from the surface, through the casing, into the annulus, to fill the space between the outer diameter of the casing and the annulus and permanently seal the casing inside the wellbore [12]. The casing is designed to isolate the drilled rock formations, to withstand high pressures and all the forces and environmental factors it will be subjected to in a wellbore and maintain integrity throughout the production of the well until it is ceased and properly abandoned. Several drill bit sizes with corresponding different casing sizes are normally utilized in the construction of a well, as shown in Fig. 1.2b-(iii), depending on the nature of the formations crossed by the drill bit and their pressures. The blowout preventer (BOP) shown in Fig. 1.2a prevents well blowouts, which can be caused by any formation fluids entering the wellbore, displacing the drilling fluid and flowing to the surface at high pressure, by closing around the drill pipe and sealing off the space between the drill pipe and the casing [13].

A schematic of a typical completed well is shown in Fig. 1.2c. The conductor casing is the first and the largest diameter casing to be cased and cement inside the well. The main function of the conductor casing is to isolate unconsolidated formations. The surface casing seals off any freshwater sands, protects fresh groundwater zones from contamination and supports the BOP. Following the surface casing is the intermediate casing, which isolates formations that may cause problems such as lost circulation zones, abnormal pressure zones and unstable shales. Finally the well is completed with a production liner that is suspended from the intermediate casing and a liner tie-back that is connected to the top of the liner hanger and extended back to the surface [12, 14]. When the well starts to produce from the hydrocarbon reservoir the liner tie-back provides the intermediate casing string protection from the pressures that will be encountered when the well is in production fluids start to flow to the surface. This is an example of one of the many possible well construction scenarios.

1.2.2 *The Downhole Environment and Drilling Challenges*

When it comes to utilizing sensors and instrumentation in extreme applications drilling a hydrocarbon well sits right at the top with space exploration [15–20]. Table 1.1 shows some typical parameters encountered in a downhole environment [21]. It is not only the high temperature, high pressure, highly acidic and corrosive environments encountered in deep wells that make drilling challenging but also the destructive conditions, such as high torque, shock and vibration, created by the drill bit grinding and penetrating through rock formations [10]. When unmonitored and unmanned, these parameters coming together in certain combinations can result in a variety of wellbore problems, which in the worst case can lead to an uncontrollable flow of hydrocarbons to the surface threatening the safety of the personnel on the

Table 1.1 Typical parameters encountered in a downhole environment [21]

Temperature	Pressure	pH	Vibration	Shock
125–230 °C	15,000–30,000 psi	2–5	30 g peak at 50–1000 Hz	1000 g, 0.5 ms, Half-sine

rig and causing major damage to not only the equipment on the rig but also to the surrounding environment and to the hydrocarbon reservoir itself. Some common problems encountered while drilling include: the drill string assembly getting stuck inside the well; axial, lateral and torsional vibrations of the drill string assembly, which can increase the wear/tear rate of the drill bit, decrease the rate of penetration and pose challenges to directional well steering systems [10, 22–27]; formation instabilities resulting in the well collapsing; and fractured formations causing the drilling fluid to flow uncontrolled into the formations, creating a situation known as lost circulation, which results in the reduction in the hydrostatic pressure exerted by the drilling fluid and the possibility of formation fluids entering the annulus and so resulting in formation fluids flowing to the surface [28–30]. Moreover, high temperature, pressure, pH, vibration and shock influence the fatigue rate and reliability of the components of the drilling assembly, especially batteries, sensors, electronics and instrumentation in logging and directional instruments [17].

Due to the aforementioned challenges, logging/directional instruments, with sensors and instrumentation designed to work in harsh downhole environments, play a crucial role in helping the driller to understand the drilling environment. This enables the drilling of wells in a safe and efficient manner and maximizing the exposure and reachability into hydrocarbon reservoirs for optimal production. Instruments must not only be able to survive the downhole conditions but also be able to acquire accurate data that helps deliver a precise representation of the downhole condition of the well so that drillers can make effective and timely decisions [31, 32]. Since the diameter of the wellbore sections decreases as depth increases drilling instruments also get smaller so components must also be small and compact enough to fit into these smaller instruments.

References

1. Business & Economics Research Advisor, Issue 5/6: Winter 2005/Spring 2006. Available online at https://www.loc.gov/rr/business/BERA/issue5/history.html#3
2. K. Ziemelis, Nature **426**, 317 (2003)
3. H.M. Johnson, in *History of Well Logging*. SPWLA 2nd Annual Logging Symposium (1961)
4. N.C. Nanda, *Seismic Data Interpretation and Evaluation for Hydrocarbon Exploration and Production—a Practitioner's Guide* (Springer, 2016)
5. H.N. Alsadi, *Seismic Hydrocarbon Exploration—2D and 3D Techniques* (Springer, 2017)
6. M. Varhaug, T. Smithson, Oilfield Rev. **27**, 63 (2015)
7. A.T. Bourgoyne Jr., K.K. Millheim, M.E. Chenevert, F.S. Young Jr., *Applied Drilling Engineering* (SPE, 1991)
8. S. Devereux, *Drilling Technology in Nontechnical Language* (PennWell, 2012)
9. A. Besson, B. Burr, S. Dillard, E. Drake, B. Ivie, C. Ivie, R. Smith, G. Watson, Oilfield Rev. **12**, 36 (2000)
10. P. Centala, V. Challa, B. Durairajan, R. Meehan, L. Paez, U. Partin, S. Segal, S. Wu, I. Garrett, B. Teggart, N. Tetley, Oilfield Rev. **23**, 4 (2011)
11. B. Mitchell, *Advanced Oilwell Drilling Engineering—Handbook and Computer Programs* (SPE, 1995)
12. E.B. Nelson, Oilfield Rev. **24**, 59 (2012)

13. R. von Flatern, Blowout Preventers, Oilfield Rev. The Defining Series (2016)
14. M. Varhaugh, Oilfield Rev. **23**, 50 (2011)
15. Z.A. Khan, B.K. Behera, V. Kumar, P. Sims, World Oil **75** (2012)
16. R. Beckwith, J. Petrol. Technol. **65**, 56 (2013)
17. C. Avant, S. Daungkaew, B.K. Behera, S. Danpanich, W. Laprabang, I.D. Santo, G. Heath, K. Osman, Z.A. Khan, J. Russell, Oilfield Rev. **24**, 4 (2012)
18. F.R. Florence, J. Burks, in *Instrumentation and Measurement Technology Conference* (IEEE, 2012), pp. 670–675
19. D.R.M. Woo, J.A.K. Yun, Y. Jun, E.W.L. Ching, F.X. Che, in *Electronics Packaging Technology Conference* (IEEE, 2014), pp. 16–21
20. K. Panetta, *Electronic Component Magazine* (2015). Available online at https://www.ecnmag.com/blog/2015/07/what-harshest-environments-components-whats-best-way-deal-those-challenges
21. C.P. Gooneratne, B. Li, T.E. Moellendick, Sensors **17**, 2384 (2017)
22. W.G. Lesso, M. Ignova, F. Zeineddine, J.M. Burks, J.B. Welch, in *SPE/IADC Drilling Conference and Exhibition*, Society of Petroleum Engineers, 2011
23. H. Santos, in *IADC/SPE Drilling Conference* (2000)
24. B. Costo, L.W. Cunningham, G.J. Martin, J. Mercado, B. Mohon, L. Xie, Oilfield Rev. **24**, 16 (2012)
25. S.M. Zamani, S.A. Hassanzadeh-Tabrizi, H. Sharifi, Eng. Fail. Anal. **59**, 605 (2016)
26. B.S. Aadnøy, K. Larsen, P.C. Berg, J. Petrol. Sci. Eng. **37**, 195 (2003)
27. B. Saldivar, I. Boussaada, H. Mounier, S. Mondie, S.I. Niculescu, IFAC Proc. Ser. **47**, 5169 (2014)
28. A. Lavrov, *Lost Circulation: Mechanisms and Solutions* (Elsevier, Oxford, 2016)
29. C.P. Gooneratne, E.S. Gomez Gonzalez, A.S. Al-Musa, H.F. Osorio, in *SPE Abu Dhabi International Petroleum Exhibition & Conference* (2017)
30. Y. Feng, K.E. Gray, J. Petrol. Sci. Eng. **152**, 511 (2017)
31. F. Florence, IEEE Instrum. Measur. Mag. **16**, 6 (2013)
32. F. Florence, F. Iversen, J. Macpherson, IEEE Instrum. Measur. Mag. **16**, 43 (2013)

Chapter 2
Instruments for Formation Evaluation

Abstract This chapter presents the two main methods of downhole formation evaluation, wireline logging (WL) and logging-while-drilling (LWD). These two methods utilize a variety of instruments to acquire a wide range of downhole measurements, which include electrical resistivity, electromagnetic propagation, spontaneous potential, gamma ray, neutron-porosity, nuclear magnetic resonance, acoustic, imaging, fluid sampling/testing and seismic, to interpret downhole formations and reservoirs. The logging instruments together with their designs and principles of measurement are also described.

2.1 Wireline Logging and Logging-While-Drilling

Formation evaluation acquires physical and chemical properties of rock formations penetrated by drilling utilizing electromagnetic, acoustic, nuclear, mechanical or other methods. This allows petroleum engineers to determine the presence or absence of commercial quantities of hydrocarbons in them, obtain information about the static and dynamic characteristics of formations and reservoirs, and acquire downhole environment parameters such as pressure, temperature and rock/fluid types. Table 2.1 shows the wide range of measurements such as electrical resistivity, spontaneous potential, electromagnetic, radioactive, nuclear magnetic resonance, acoustic, spectrometric and seismic available to record hydrocarbon potential properties. Porosity is the percentage of void space in rock formations and permeability is the degree of interconnection between void spaces that allow fluids to flow through the voids. Lithology refers to identifying macroscopic physical properties of rock formations, such as colour, texture, grain size or composition, and classifying them into different rock layers such as sandstone, limestone, claystone, shale, anhydrite etc. Mineralogy is the study of the chemistry, crystal structure as well as physical properties, such as density, of minerals that exist in rock formations. The sizes and shapes of pore geometry within a rock formation play an important role in identifying and optimizing drilling and completion fluids to control fluid loss and invasion, respectively. These properties along with the

C. P. Gooneratne et al., *Instruments, Measurement Principles and Communication Technologies for Downhole Drilling Environments*, Smart Sensors, Measurement and Instrumentation 32, https://doi.org/10.1007/978-3-030-04900-3_2

Table 2.1 Logging measurements used to determine reservoir properties [7]

Measurements	Resistivity	Porosity	Lithology	Mineralogy	Saturation	Pore geometry	Permeability	Fluid properties	Geomechanical properties	Geologic structure	Geologic bedding
1. Electrical resistivity											
i) Laterolog	Blue		Orange		Blue	Orange	Orange	Orange		Orange	
ii) Microlaterolog	Blue		Orange		Blue			Orange		Orange	
iii) Induction	Blue		Orange		Blue		Orange	Orange		Blue	Green
2. Spontaneous potential			Orange				Orange	Orange		Orange	
3. Electromagnetic propagation	Blue	Blue	Orange		Blue	Blue	Orange	Blue		Orange	Blue
4. Nuclear											
i) Gamma ray density		Blue	Green	Green				Blue	Green	Orange	
ii) Neutron porosity		Blue	Green	Green				Blue		Orange	
iii) Natural radioactivity			Blue	Green						Orange	
5. Induced gamma ray spectrometry		Green	Blue	Blue			Orange		Green	Green	
6. Nuclear magnetic resonance		Blue	Green	Orange	Blue	Blue	Blue	Blue	Orange	Orange	
7. Acoustic		Blue	Blue	Green	Green	Blue	Blue	Blue	Blue	Orange	
8. Dipmeter and imaging	Blue	Orange	Orange	Orange		Blue		Orange	Blue	Blue	Blue
9. Formation testing and sampling											
Fluids sampling					Green			Blue		Orange	Orange
Fluids pressure testing					Green		Blue	Blue		Orange	Orange
10. Seismic		Green	Orange		Green			Green	Green	Blue	

- Blue: Measurement provides direct information about the reservoir property
- Green: Measurement is influenced or is sensitive to the reservoir property
- Orange: Measurement contributes to understanding the reservoir property

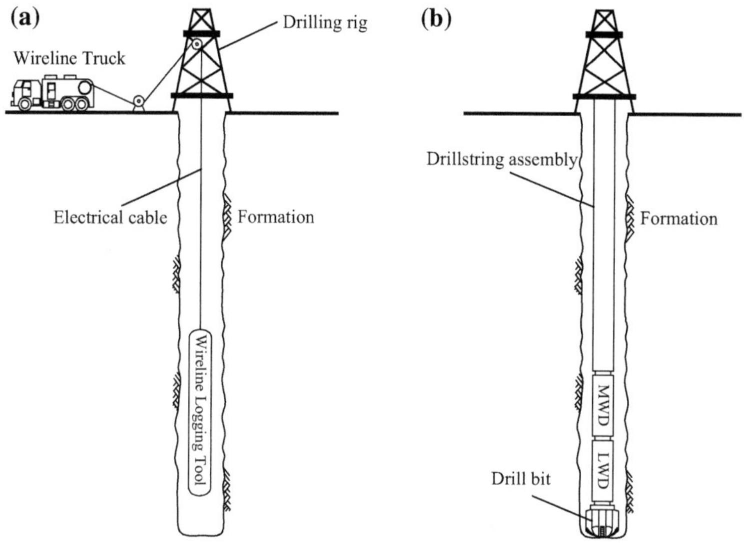

Fig. 2.1 **a** A wireline operation where the logging instruments are lowered into the well by an electrical cable connected to a truck. **b** A measurement and logging while drilling operation where real-time data is obtained while drilling a well

geomechanical and geologic properties of formations provides the driller an indication of how easy/difficult it would be to drill through formations and the problems [1–10]. This data combined with temperature, pressure and the analysis of wellbore fluids, rock cuttings and rock samples (when available) can then be used to optimize drilling programs.

The earliest and the most established method of obtaining measurements downhole is by an operation called wireline (WL) logging [7, 11–16]. Figure 2.1a shows a WL logging truck, which has a winch containing 25,000–35,000 ft of electrical cable, a logging unit with instruments connected to the cable, and a surface unit to power the logging unit, acquire data and record them as logs. The logging instrument, also known as a logging tool, is lowered into the well by unwinding the cable around the winch under tension. The logging tool consists of two components, one called the sonde, which houses the formation evaluation sensors, and another called the cartridge, which contains the electronics, the signal processing and data processing units and communication modules. Many WL logging tools have sondes for different formation evaluation instruments and their corresponding cartridges connected together as one tool. The WL logging tool is run inside the well and measurements are obtained as a function of depth during the ascent into, and the descent out of, the well. The depth is measured with a calibrated measuring wheel system on which the cable rides on.

The data from the logging unit is transferred through the cable to the surface. A data acquisition and processing unit in the truck then performs data analysis and signal processing on the raw data and transfers the processed data through a satellite communications link to a central location for data interpretation by petroleum engineers.

While WL tools have been utilized for many decades to obtain downhole formation data, it cannot be run during drilling. Therefore, the drilling assembly has to be pulled out of the well first before lowering the WL tool. The logging operation can also be time consuming, depending on the well depth and resolution of the data; generally the logging speed has to be slowed down for acquiring high resolution data. Once the logging measurements are obtained the WL tool then has to be pulled out of the well before drilling can resume. Moreover, during a WL operation, drilling fluid in the well typically remains static and the wellbore condition has to be mechanically stable. If several logging runs are required, a dedicated pipe trip to clear any obstructions in the well is required in between the logging runs. Finally, there is always the risk of a WL logging tool getting stuck in the hole, which can significantly add to the cost of drilling a well. All these factors make a WL logging a riskier, time-consuming and overall expensive operation [17].

Ceasing a drilling operation for WL logging may not be optimal due to economic and operational reasons. Moreover, drilling a well does not provide real-time knowledge of the rock formations being penetrated and the downhole environment. For a long time, the obstacle for formation evaluation while drilling was data communication between the downhole instruments and the surface. The emergence of a telemetry method in the 1970s based on changes in drilling fluid flow pressure or pressure waves paved the way for measurement-while-drilling (MWD) and logging-while-drilling (LWD) tools. MWD, as discussed in Chap. 3, is related to navigating a well and obtaining drilling dynamics data. Today, LWD tools are able to acquire, in real-time, most formation parameters obtained by WL logging, and in some situations even preferred [18–20]. LWD instruments for formation testing and sampling and micro-resistivity are still in the development stage but have shown to be feasible for use in drilling environments. On the other hand, there are no commercially available WL tools that obtain azimuthal density measurements. The MWD/LWD tools, as shown in Fig. 2.1b, obtain real-time data while drilling but also have the capability to store high resolution data in memory, which can be retrieved and compared with real-time data for accuracy.

LWD tools, as the name suggests, obtain downhole measurements, also known as 'logs', while drilling [21–24]. The type of measurements and the formation parameters evaluated are the same as in WL logging, with some differences in resolution, coverage and quality [17, 25–36]. While LWD tools are larger and heavier than WL tools and data communication slower, they provide real-time data while drilling, which allows drilling engineers to immediately respond to well control problems and revise drilling programs. Wells drilled can be vertical or directional, where a directional well is drilled by deviating from its vertical path at a given angle. A well is considered horizontal if the angle of deviation from the vertical is more than 85°. Drilling directional/horizontal wells is covered in Chap. 3. WL is generally restricted up to an inclination of 65° while MWD/LWD tools can be utilized for any conventional directional/horizontal well that is drilled. Moreover,

LWD tools obtain azimuthal measurements from around the circumference of the wellbore due to the rotation of the drilling assembly, and are better suited for logging in deviated wells than WL tools. LWD tools utilize MWD data along with several algorithms and statistical analyses to compensate for standoff, lamination and the wellbore being roughly parallel to the rock formations.

Interpretation of early well logs were mostly qualitative in nature whereas most current logs are quantitative with more refined qualitative aspects, enabling detailed analysis and interpretation of data trends and patterns. The measurements in Table 2.1 do not always provide an accurate interpretation of formation/reservoir properties when used as stand-alone measurements. Therefore, a combination of measurement principles are typically utilized to gain more accurate insights. Furthermore, information such as volume, thickness, distribution and extent of hydrocarbons in a reservoir cannot be directly measured and are therefore, derived and estimated from various combinations of these measurements using a number of assumptions. The WL and LWD tools come in combo packages, where the drilling engineer has the option of choosing the specific instruments required for a given well section. An example of a common combination of logging measurements are shown in Fig. 2.2.

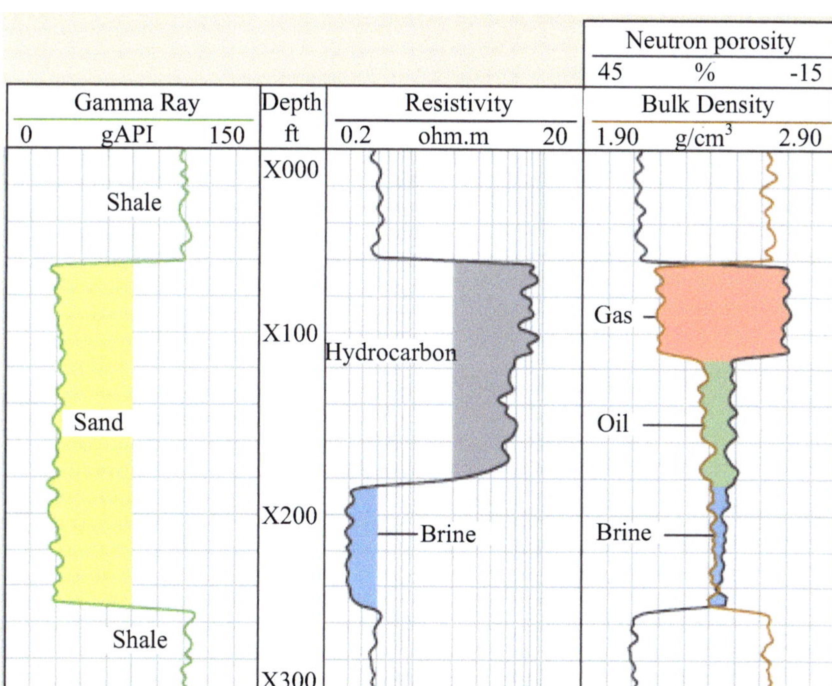

Fig. 2.2 Gamma ray, resistivity, neutron porosity and bulk density measurements used to identify sand and shale formations as well as the presence of gas, oil and brine along a depth of 300 ft [7]

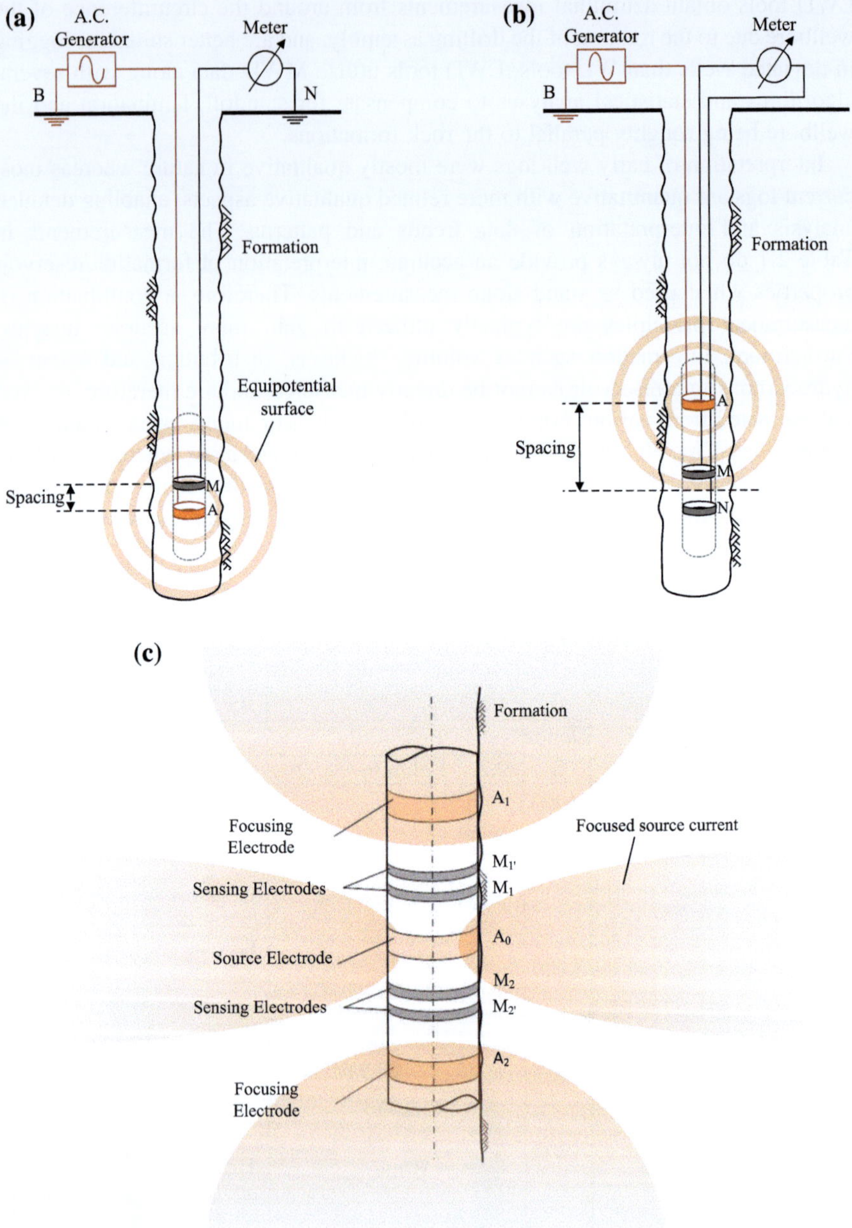
(a)
A.C. Generator
Meter
B
N
Formation
Equipotential surface
Spacing
M
A
(b)
A.C. Generator
Meter
B
Formation
A
Spacing
M
N
(c)
Formation
A_1
Focusing Electrode
Focused source current
$M_{1'}$
Sensing Electrodes
M_1
A_0
Source Electrode
M_2
Sensing Electrodes
$M_{2'}$
A_2
Focusing Electrode

◂**Fig. 2.3** Electrical resistivity measurement instruments. **a** Normal log instrument with current flowing from electrode A inside the well to B outside creating equipotential surfaces, and measurement of voltage, which is a function of resistivity, between electrodes M inside the well and N outside. **b** Laterolog instrument where measurement electrodes M and N are both placed inside the well and below electrode A to measure the voltage gradient leading to a deeper resistivity measurement. **c** Focused laterolog instrument where the emitted current by source electrode A_0 is focused by A_1 and A_2 into a thin disk region of the formation. Measuring the voltage between the upper, M_1 and $M_{1'}$ or lower, M_2 and $M_{2'}$, set of sensing electrodes and the voltage at a distance far from the sensing electrodes, and knowing the magnitude of the source, allows the calculation of the resistivity of the formation

Most of the measurements are influenced by drilling fluid characteristics such as salinity, density, solids concentrations etc., together with temperature, pressure, size and rugosity of the wellbore, tool alignment, logging speed, as well as the lithology, pore size, type of fluid in the pores and the geologic structure and geometry of the rock formation. An in-depth analysis of the methods utilized to correlate measurements acquired by logging instruments to downhole formation properties and how different variables and combinations of variables influence the interpretation of logging measurements are beyond the scope of this book. Therefore, this chapter focuses on the main formation evaluation instruments and their measurement principles and configurations, which form the basis for all the successive instruments and the current instruments used in the drilling industry, and how these measurements are utilized to infer knowledge about rock formations and the downhole environment. A number of leading references provide the interested reader the theoretical analysis and principles of interpretation of the measurements acquired by the instruments, along with their relationships to downhole formation properties [1–3, 6–10, 12–17].

2.2 Principles of Measurement and Interpretation

2.2.1 Electrical Resistivity

Resistivity, the reciprocal of conductivity, measures the degree to which a formation resists the flow of electric current and is measured in units of ohm-m. Resistivity logs in principle can be utilized to distinguish between water-bearing and hydrocarbon zones, since water found in formations have salt ions and are therefore conductive and have low resistivity, whereas hydrocarbons being non-conductive have a relatively higher resistivity [1–3, 6–10, 12–17]. The measured resistivity values increase as the pores in a formation become more saturated with hydrocarbons. When combined with porosity and formation water resistivity, resistivity measurements are used to obtain the water saturation in a reservoir, which in turn is used to evaluate the production potential of the reservoir. Modern instruments are able to measure resistance of the section near the wellbore, the section of a formation invaded by drilling fluids called the flushed zone, and the uninvaded section by shallow, medium and deep resistivity measurements.

There are three main methods of resistivity measurements as shown in Table 2.1; (i) laterolog (ii) microlaterolog and (iii) induction. Figure 2.3 shows how laterolog type measurement instruments evolved from the initial normal resistivity measurement instrument. Figure 2.3a shows an electrode A inside a well and another electrode B at the surface. When an alternating current is passed from electrode A to B, the current will flow radially with equipotential surfaces around electrode A. Another electrode M introduced into the well, where M is placed closed enough to A for the current to be constant, will lie on an equipotential surface generated by A. When a voltmeter is placed between M and another electrode N at the surface, a voltage, which is a function of the resistivity between A and M, can be measured. This normal arrangement of electrodes, where A and B form the current generator circuit and M and N formed the sensing circuit, were the first instruments used for resistivity measurements in wells [37–41]. In a laterolog instrument, as shown in Fig. 2.3b, electrode N is placed close to electrode M and they are located several feet below A, while electrode B is still at the surface. This configuration measures the voltage gradient between M and N and provides a deeper resistivity measurement than normal tools by increasing the spacing between electrode A and electrodes M and N [3, 38–43]. The focused laterolog instrument shown in Fig. 2.3c forms the basis for the currently available WL laterolog resistivity instruments. There are 7 electrodes, where A_0 is the current emitting source electrode, A_1 and A_2 are focusing electrodes, M_1, $M_{1'}$ are upper sensing electrodes and M_2, $M_{2'}$ are lower sensing electrodes. The focusing electrodes emit a current to focus the source current in a thin disk into the formation [3, 43–47]. Therefore, M_1 and $M_{1'}$ and M_2 and $M_{2'}$ measure the same potential difference and this aids the focusing currents in A_1 and A_2 to automatically adjust to keep the source current focused into the formation. Measuring the voltage between the upper or lower set of sensing electrodes and the voltage at a distance far from the sensing electrodes, and knowing the magnitude of the source, allows the calculation of the resistivity of the formation.

These early focused laterolog devices were followed by dual laterolog instruments that allowed both a shallow and a deep penetration into the formation [48–54]. The dual laterolog has 9 electrodes, with two extra upper and lower focusing electrodes. In the shallow operating mode the extra upper and lower electrodes act as ground electrodes to the current emanating from the focusing electrodes, so that the current from the source electrode penetrates far less into the formation. In the deep operating mode the extra upper and lower electrodes emit a current so that the current from the source electrode is penetrated deeper into the formation. Spherically focused laterolog instruments have an electrode arrangement where the current is focused in a quasi-spherical pattern into the formation [48, 55]. The laterolog instruments have a dynamic range from 0.2 to 20,000 Ωm. Azimuthal dual laterolog and high resolution azimuthal laterolog instruments have a set of segmented azimuthal electrodes that measure shallow and deep resistivity variations around a wellbore [56–60]. The most current laterolog instrument is the array laterolog instrument, which have segmented focusing electrodes and the ability to measure resistivity at multiple depths by working at different frequencies [61–66].

These instruments use a two dimensional (2D) earth model and inversion scheme to obtain more accurate resistivity measurements.

Microlaterolog instruments are a smaller version of the laterolog instruments previously mentioned. These instruments have electrodes on pads that are pressed against the surface of the wellbore for measurements [49, 67–72]. The distances between the electrodes are much smaller than in laterolog instruments so they only penetrate a very shallow section of the formation. A typical microlaterolog instrument has a button electrode as the source current surrounded by two rings of sensing electrodes and a focusing electrode ring, which is farthest away from the source current electrode. The spacing between the electrodes is no more than 1 inch. Variations of the microlaterolog include proximity and micro-spherically focused instruments [48]. Microlaterolog-based techniques are seldom used as quantitative measurements and serve mainly as the basis for electrical imaging tools.

Induction instruments are used in wells that have moderate to highly resistive drilling fluids [42, 73, 74]. Figure 2.4 shows an induction instrument with a transmitter and a receiver coil. A low frequency alternating current, typically between 10 and 40 kHz to avoid significant dielectric effects, applied to the transmitter generates a magnetic field inducing eddy currents that flow around the rock formation. These eddy currents produce a secondary magnetic field that induces a current in the receiver. The eddy currents and the magnitude of the secondary magnetic field are proportional to the conductivity of the formation, so a higher signal at the receiver indicates a more conductive formation. Current induction instruments built on this measurement principle have particular arrays of transmitter and receiver coils to compensate for the skin-effect and for investigating specific depths of the formation, and focusing coils to reduce direct coupling between transmitters and receivers [75–80]. A lot of research and modeling have been done to develop algorithms to optimize induction logging instruments in inhomogenous anisotropic formations, deviated wells and in the presence of natural fractures [81–88].

Current LWD resistivity instruments generally use electromagnetic fields with frequencies between 100 kHz and 2 MHz to measure propagation resistivity in formations [1, 38, 89–91]. Figure 2.5 shows a schematic of a LWD propagation resistivity instrument with two receivers sandwiched between a pair of upper and lower transmitters equally spaced on either side of the receiver pair. Each transmitter alternately transmits an electromagnetic field into the formation and the attenuations and phase-shifts measured between the two receivers are averaged and transformed into resistivity readings; T_1 to R_1 and R_2, T_2 to R_1 and R_2, T_3 to R_1 and R_2, T_4 to R_1 and R_2. The phase-shift provides a resistivity measurement at the flushed zone of the formation while the attenuation provides a resistivity measurement at the uninvaded zone of the formation. The configuration and the equidistance of the transmitter and receiver pairs ensure a compensated resistivity measurement [1, 38, 90, 92, 93]. The numbers and orientation of, and spacing distances between, transmitters and receivers and the transmission frequencies may vary between instruments, but the principle of measurement remains the same [38, 94–101].

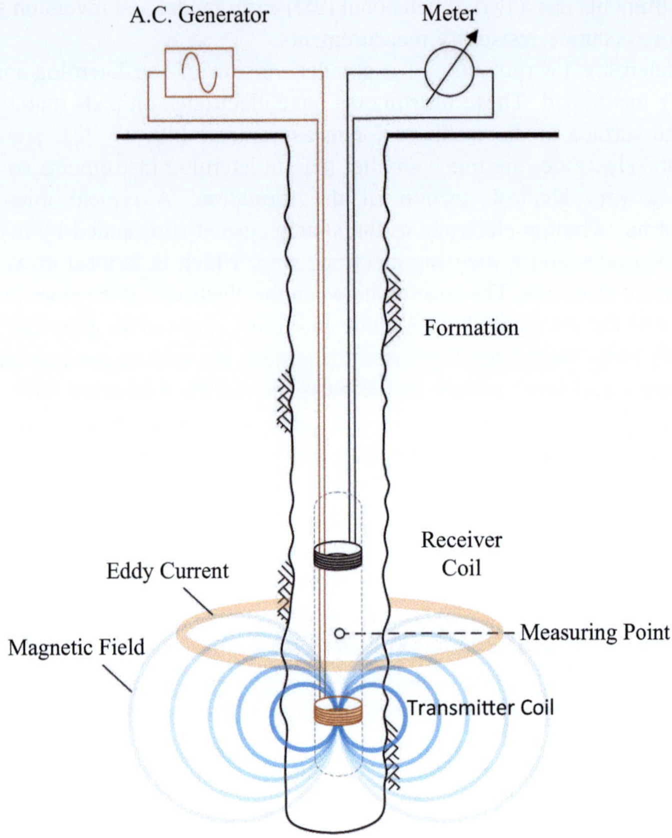

Fig. 2.4 Induction logging instrument that induces eddy currents in rock formations by transmitting a high frequency magnetic field. A receiver coil detects the secondary magnetic field from the eddy currents, which is proportional to the conductivity of the rock formation

2.2.2 *Electromagnetic Propagation*

Electromagnetic propagation instruments measure the dielectric properties of rock formations, such as the dielectric constant and the conductivity [93, 102–105]. The dielectric constant is the ratio of the permittivity of the rock formation to the permittivity of free space, where the permittivity of a rock is a measure of its degree of resistance to allow the development of an electric field. Dielectric propagation measurements are performed between 20 and 200 MHz while similar measurements performed at frequencies above 200 MHz up to 1 GHz are known as electromagnetic propagation logs. While low frequency dielectric propagation logs have the advantage of larger depths of investigation, reaching therefore into the non-invaded zone of a formation, dielectric propagation logs are influenced more by salinity than electromagnetic propagation logs making their interpretation more

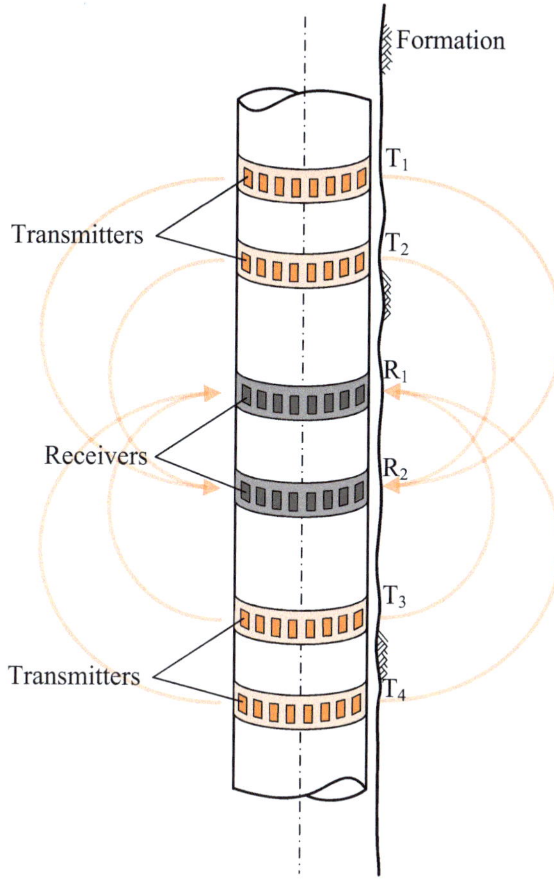

Fig. 2.5 LWD resistivity instrument where the attenuations and phase-shifts of the electromagnetic fields transmitted by the transmitters are measured between the two receivers and averaged and transformed into resistivity readings

complex. Dielectric measurement instruments, first developed in the 1970s, are used to determine water saturation in oil bearing formations when the formation water has low or no salinity [38, 106–110]. While resistivity measurements are able to differentiate between hydrocarbons and saline water in formation pores it fails to accurately make this distinction when pores are filled with fresh water [1, 2, 7, 12]. Porosities are usually derived from water content and in general, dielectric permittivity increases with increasing water content and increasing porosity of brine-saturated rock.

Figure 2.6 shows an electromagnetic propagation pad type device with two transmitters and two receivers on the pads. The transmitters transmit electromagnetic waves that propagate through a formation and are detected by the receivers. The transmitters and receivers are placed on the pad to minimize the effects of borehole rugosity, instrument misalignment and weak pad contact on the measurements. The basic measurements at the receiver are the phase shift and the attenuation of the transmitted electromagnetic waves. As with LWD resistivity

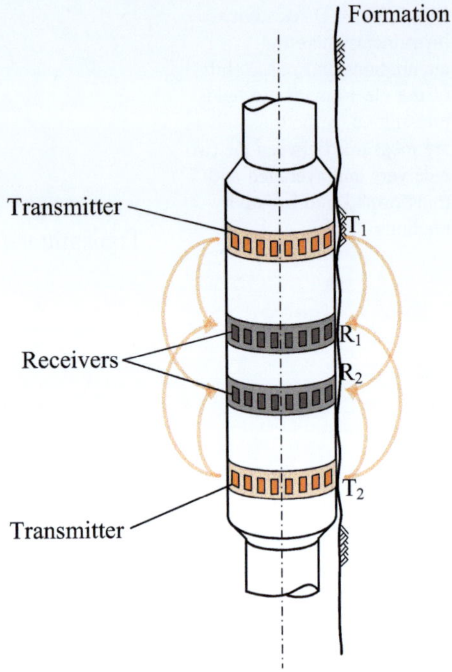

Fig. 2.6 Electromagnetic propagation instrument that measures the dielectric properties of rock formations. The transmitters transmit an electromagnetic field that move through a formation and the attenuation and phase-shift of the field detected at the receivers provides an indication of the dielectric properties of rock formations

propagation instruments dielectric logs have been recorded by many different variations of electromagnetic propagation instruments with different transmitter and receiver configurations [38, 111–113].

2.2.3 *Spontaneous Potential*

The spontaneous/self-potential (SP) log records the voltage between a movable electrode in the well and another electrode fixed at the surface, as shown in Fig. 2.7. The voltage between the two electrodes, generally in the order of mV, varies according to the physical-chemical interactions between the drilling fluids, formation fluids and cuttings in the well since the electrochemical and electrokinetic potentials in the formations cause electric currents to flow in the drilling fluid [114–116]. Therefore, SP logs can only be recorded in conducting drilling fluids. If the ionic concentration of the drilling fluid is less than the ionic concentration of the formation fluid, the SP log will be negative, and in the opposite case, positive. Since the voltage will depend on the exchange of ions between the drilling and formation fluids, rock formations such as clay and shale, characterized by low porosity and permeability, will produce a 'flat' line that can be used as a base reference line in SP logs [117, 118]. Also, SP logs show a reduced response when the pores have hydrocarbons so they may be used to diagnose rocks for hydrocarbon potential.

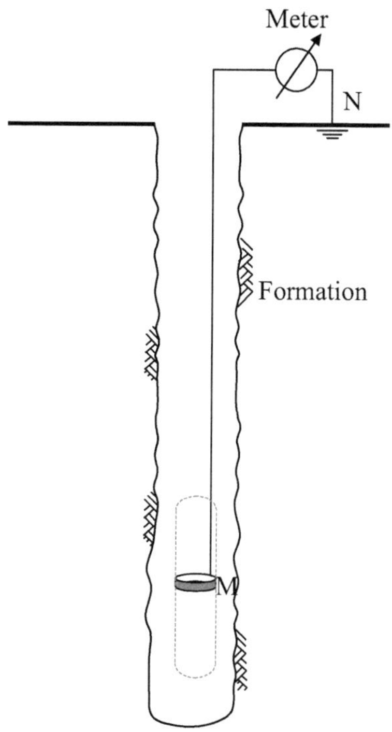

Fig. 2.7 Instrument to measure the spontaneous potential of the well. The voltage measured between electrodes M and N is influenced by the type of drilling fluids, rock formation fluids and the rock cuttings in the well

SP instruments are generally run with gamma ray instruments to provide a more accurate indication of permeability, porosity and shale volume [1, 2, 6, 8–10, 12, 119].

2.2.4 Gamma Ray

The first gamma ray (GR) logs run in the 1930s were an advancement over SP logs since they were not dependent on the conductivity of drilling or formation fluids [120]. The basis for GR measurements is the naturally occurring radioactive elements, which include uranium, thorium, potassium, radium and radon, in rock formations [121, 122]. GR detection instruments measure this natural radioactivity emanating from the formations using sensitive detectors and associated electronics that count the number of gamma rays per unit time; GR detection instruments are passive [123–125]. Figure 2.8 shows a GR detection instrument with a scintillation detector containing a scintillation crystal, which is generally sodium iodide but can be any material with similar properties, and a photomultiplier. When a gamma ray strikes the scintillation crystal a single photon of light is released, where its intensity depends on the energy of the GR. The photomultiplier detects this photon and outputs an electrical pulse with an amplitude proportional to the intensity of the

Fig. 2.8 Passive gamma ray instrument that measures the dose of gamma rays in rock formations by a scintillation counter and photomultiplier system

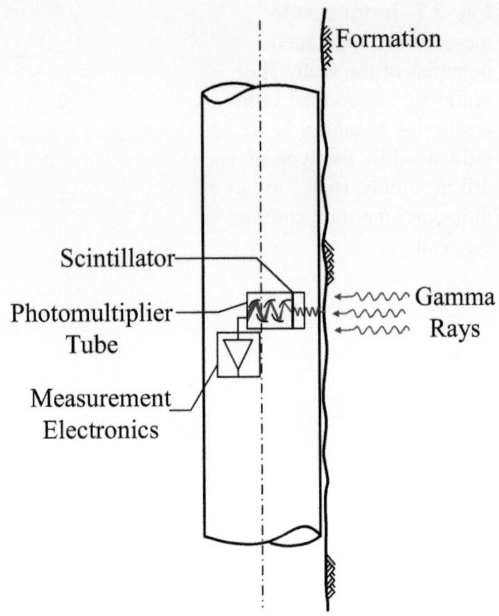

scintillation. Therefore, the scintillation counter and the photomultiplier system can provide the count rate and the energy distribution of the natural radiation emanating from a rock formation. Interpretation of GR logs provides information on the lithology and mineralogy of rock formations that reflect their mechanical and petrophysical properties [1, 2, 6, 8–12, 125]. For example, radioactive elements tend to concentrate in clay and sylt-rich formations so GR logs show high values for these formations while clean formations being less radioactive show relatively lower values. Gamma rays emanated randomly from formations fluctuate with time. Depending on the efficiency of the utilized detector and on the objective of the measurement GR detection instruments must be run at compatible speeds to acquire enough 'events', in order to obtain high quality data. Gamma rays can be measured through steel casing so it can be run in both open hole and cased hole sections. GR instruments frequently complement SP logs and are used as a substitute for SP logs if wells are drilled with saline or oil-based drilling fluids or air. GR tools are also usually used in combination with other tools for correlation.

Figure 2.9 shows a schematic of a basic GR density instrument, which invariably feature at least one GR source, almost always caesium-137 (Cs-137), and one GR detector. Uncompensated density measurements are derived from the radiation reaching the GR detector(s), which originated from the radiation emanated from the source and was scattered by the rock formation. Compensated formation density instruments appeared in the 1960s to correct for wellbore effects. GR density instruments based on the photoelectric effect first appeared in the 1980s [126–130]. Modern instruments typically consist of a focused radioactive source, typically caesium-137 or cobalt-60, which emits gamma rays in a medium energy region between 0.2 and 2 meV, and a near and a far detector, where the detectors are very similar to the gamma ray detectors explained before. The source and the detectors

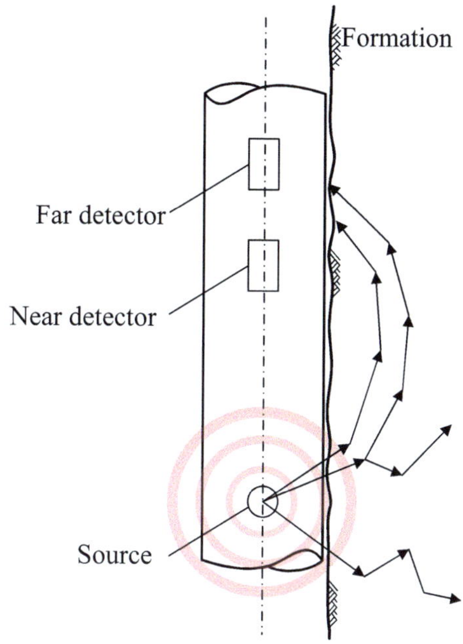

Fig. 2.9 Active gamma ray instrument with a source to emit gamma rays into a rock formation and near and far detectors that measure the attenuation of the gamma rays after travelling through the formation. The gamma ray signal measured at the detectors provides an indication of the density of the formation

are shielded so that the gamma rays will go into the drilling fluid and formation and not directly from the source to the detectors. Having two detectors compensate for any effects the drilling fluid might have on the measurements.

The gamma rays emitted by the source enter the formation and act as high-velocity particles colliding with the electrons in the formation and losing energy in the process. After each collision a gamma ray loses energy in a step-wise manner to the electron and continues with diminished energy in an interaction known as Compton scattering. The attenuated flux of gamma rays eventually reach the short and long range detectors, where the magnitude of attenuation depends on the density of electrons in the formation. Since there are natural gamma rays emitted from the formation a background gamma ray count is performed to calibrate the gamma ray density instrument. Formations with a high bulk density have a higher density of electrons and this translates into lower gamma ray rates at the detectors. Formations with lower bulk densities have low electron densities, resulting in less GR attenuation and therefore higher gamma ray counts at the detectors. The bulk density of a formation is comprised of the density of the rock material and the fluids in its pores. Therefore, GR density logs primarily provide information on porosity but also on lithology, mineralogy and gas-bearing formations [1, 3, 6, 8–12, 131–135]. The lithology and mineralogy are obtained by so-called litho-density instruments that discriminate incoming radiation energy levels and obtain both Compton-region scattering and the photoelectric absorption spectra; photoelectric absorption generates low energy gamma rays, measuring less than 0.2 meV. It is originated by the interaction between electrons and low energy gamma rays. GR density logs also aid in the determination of mechanical properties of rock formations.

Spectral GR tools have the added sophistication of being able to sort the electrical pulses from the photomultiplier, which is proportional to the brightness of the light pulse produced by the detector, into separate bins based on their amplitude [136, 137]. This method of segmentation, known as pulse-height spectrum analysis, is then used to create a full energy spectrum of the incoming gamma ray from the formations. The spectra of the naturally occurring GR elements is typically used to generate information about the KUT (Potassium-Uranium-Thorium) concentration of rock. K only produces a gamma ray whereas U and T initially produces a gamma ray that then further decays into other elements called daughters; the daughter elements are also radioactive, producing further gamma rays [138–140]. Obtaining KUT concentrations from a GR log enables better identification of clays and detection of moved water by the count rate of U. Different amounts of potassium, thorium and uranium are associated with different amounts of shale in formations and different ratios of these elements can provide further information on rock formations [140–144].

2.2.5 *Neutron-Porosity*

Neutron-porosity instruments, which first appeared in the 1940s, exploit the presence of hydrogen atoms in formations to determine porosity [145]. The amount of hydrogen in a formation can be correlated to the fluid, water and/or hydrocarbon content in formation pores. Figure 2.10 shows a neutron-porosity instrument with a neutron source and a near and a far detector. The neutron source, typically

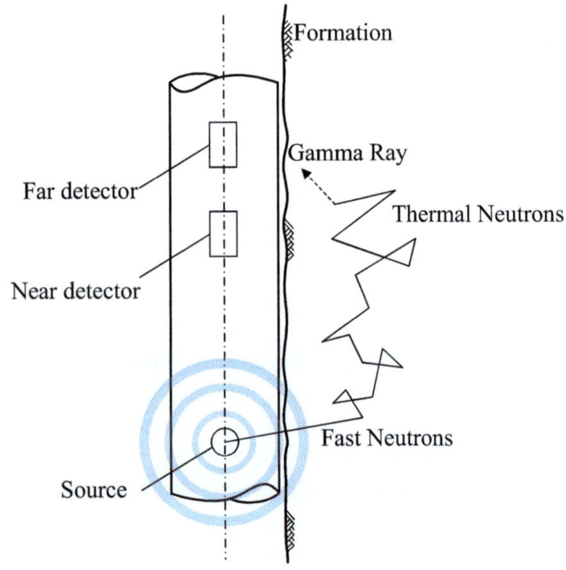

Fig. 2.10 Neutron porosity instrument to determine the porosity of a rock formation. The source bombards neutrons into the formation, which collides with nuclei in the formation and scatter, losing energy and producing high energy gamma rays. The amount of gamma rays and thermal neutrons measured at the near and far detectors provides an indication of the porosity of the formation

carbon-12, a mixture of an alpha radiation source such as radium, plutonium or americium and beryllium-9, is used to continuously bombard the formation with high energy neutrons. The neutrons collide with nuclei in the formation and scatter, losing energy and producing high energy gamma rays in the process. The gamma rays and the slowed down thermal neutrons are detected by the near and far detectors, which makes measurements less sensitive to downhole environmental effects than having a single detector [146–151].

The energy loss of a neutron per collision is related to the mass of the nucleus it collides with. When a neutron collides with a nucleus that has a higher mass it does not lose too much energy and slow down significantly but when it collides with a nucleus that has an equal mass it loses much more energy and slows down considerably. The nucleus of the atom it collides with also gains energy during this collision process so the energy transfer is also more efficient when the neutron and nucleus have similar masses. Hydrogen has a similar mass to neutrons, so in formations with high hydrogen content a neutron is slowed down to its thermal velocity very quickly and captured by hydrogen nuclei. Therefore, the count rate of thermal electrons or gamma rays will be low and the rock formation porosity high [1, 2, 6, 8–12]. However, in a formation with a low hydrogen content a neutron requires a much longer path to slow down and get absorbed, which results in a high count rate of slow neutrons or gamma rays and low rock formation porosity. The neutron-porosity instruments are corrected for chlorine, since formation and/or drilling fluids may also contain chlorine, which like hydrogen contributes significantly to neutron absorption.

Induced gamma spectrometry, based on the principle that thermal neutrons captured by atomic nuclei results in the emission of gamma rays, has been used for nearly 40 years as an instrument to obtain measurements of elements such as hydrogen, silicon, calcium and chlorine, to aid in the mineralogical analysis of rock formations [152–155]. However, it must be noted that many types of clay have a high amount of hydrogen and may significantly influence the neutron logs to show a higher porosity than what is present in a formation. The emission energies of the gamma rays represent the specific elements responsible for capture resulting in a GR spectrum. Moreover, the count rate for each signature in the spectrum correlates to the proportion of that element present in the rock formation. These instruments used in induced gamma ray spectrometry, commonly called pulsed neutron tools, are similar to Fig. 2.10 but have accelerator type pulsed neutron generators emitting neutrons at 14 meV of energy, with a similar elemental mix as in neutron-porosity instruments, to bombard the formation and near, far and sometimes deep gamma ray detectors [156–162].

2.2.6 Nuclear Magnetic Resonance

Nuclear magnetic resonance (NMR) offer a non-radioactive option for determining the porosity and permeability of rock formations by magnetically manipulating and measuring the NMR signals from the hydrogen nuclei in rock formation fluids [1, 3,

6, 8–12]. Since the introduction of the first NMR instrument in the 1960s [163–165] and the first 'inside out' instrument in the 1989 [166–168] a number of studies were done in the 1990s to investigate the relationship between multi-exponential relaxation of hydrogen nuclei and complex networks of pore spaces and their diffusion regimes [169–173]. These initial studies formed the foundation for understanding pore size distributions of different rock formation media and led to 2D NMR maps based on diffusion and relaxation [174, 175]. Research efforts in the 2000s led to single-sided NMR instruments with strong, homogenous magnetic fields to polarize the hydrogen nuclei in rock formations and also paved the way for unlocking the fluid properties from the NMR data to determine the different types of fluids present in rock formation pores [176–178]. Nowadays NMR instruments are able to obtain 3D, correlation of fluid properties for nuclei relaxation times and diffusion, and 4D, inclusion of the radial distance from the formation wall, NMR maps for accurate profiling of rock formations [179].

Figure 2.11a shows a general schematic of an NMR instrument. NMR instruments are 'inside out' because the geometry of the magnets and the transmitter/receiver coils are inverse to the geometry of standard laboratory and medical NMR instruments. The magnets produce a radially symmetric, high strength static magnetic field that is projected out to the rock formations and orients the hydrogen nuclei to be in the same direction as the magnetic field as shown in Fig. 2.11a. The time taken to fully polarize hydrogen nuclei is known as the T_I time. A 90° radio-frequency (RF) magnetic field, applied by the transceiver as a series of short pulses, is then applied to tip the nuclei away from the static magnetic field and to precession motion, as shown in Fig. 2.11a inset. The spins have different angular momentums, so a 180° pulse at the same frequency as the RF pulses flips the direction of the net spin so that the spins refocus at a certain time. Strong echo signals are produced between the train of 180° pulses and these signals decay exponentially with a relaxation time of T_{II}. The echo signals induce a voltage in the transceiver and the voltage amplitude gives an indication of the amount of hydrogen nuclei present in a formation. This technique is known as the pulse-echo technique [180, 181]. The rock formation porosity can be determined by the extrapolation of the first echo with accurate exponential fitting [182]. The relaxation time T_{II} also provides information on pore size, where longer T_{II} values indicate larger pores due to less frequent collisions between nuclei and the walls of a fluid filled pore and shorter T_{II} values indicate smaller pores, as shown in Fig. 2.11b. The pore size is used to estimate the permeability of rock formations since the permeability is proportional to the square of the diameter of the pore [183–185]. The pore size distribution, another parameter derived from NMR logs, is analyzed to obtain a geological interpretation of the rock formation. NMR measurements are also used to characterize different formation fluids based on their viscosities [179, 182, 186].

The earliest NMR tools in the 1990s were run inside the well using WL and LWD-NMR tools started appearing in the early 2000s [187–189]. LWD-NMR is gradually replacing WL tools and recent research into NMR has aided the LWD-NMR tools to provide more accurate and better characterization of rock

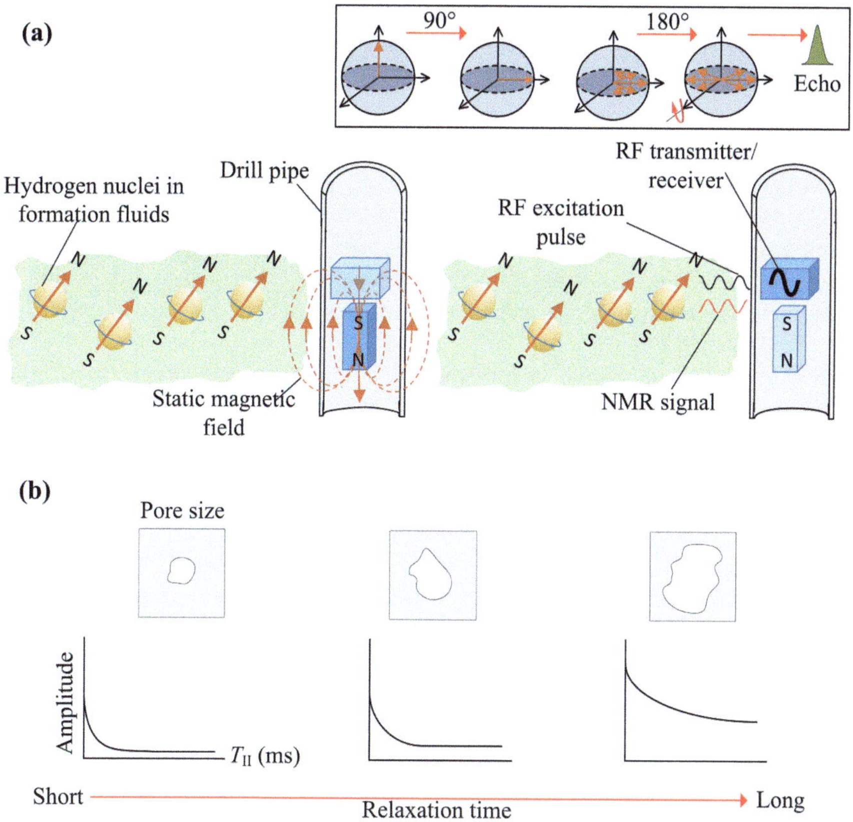

Fig. 2.11 **a** NMR logging setup, where the magnets produce a static magnetic field to orient the hydrogen atoms in the rock formations to the same direction as the field. Then, an 90° RF excitation pulse produced by an RF transceiver tips the nuclei away from the field and onto precession motion and a further 180° pulse ensures the spins refocus at a certain time (see inset). The polarization and decay of a nucleus results in echo signals, which are detected by the RF transceiver. **b** In fully water saturated pores larger pore sizes indicate long relaxation times and smaller pores short relaxation times

formations [190–193]. The 'inside out' geometry of the instrument and real-time data acquisition while drilling introduces several complexities such as the need for radially symmetric, high strength static magnetic fields and high frequency RF magnetic fields to ensure sufficient penetration through the wellbore and into geological formations, borehole size and resultant limits on tool real estate, tool eccentricity inside the wellbore, drilling fluids, downhole environmental parameters such as, temperature, pressure, pH etc., along with drilling conditions such as shock and vibration. Different techniques are used by commercially available LWD-NMR tools to address these issues and compensate for their effects on the measurement results [179, 187–189, 194–197].

2.2.7 Acoustic

Measurements obtained by acoustic instruments are based on the attenuation and the time taken for sound waves, transmitted by an acoustic source, to go through drilling fluids, rock formations and back through the drilling fluids to a detector. Acoustic logs provide direct measurements on, or are used as correlation logs, across most of the formation parameters of interest shown in Fig. 2.12. Over the years a tremendous amount of research has been performed on acoustic logging methods [198–206]. It is important to note that the frequencies of operation of acoustic instruments are higher than the frequencies used for seismic analysis that is discussed later in this section.

Figure 2.12 shows a sonic instrument with a transmitter, either a piezoelectric or electromechanical device that converts an electrical signal to a pressure wave, and a near and a far detector, which are again piezoelectric devices but converts the received pressure waves to electrical signals. The sonic waveform, usually around 20 kHz, transmitted by the transmitter arrives at the near and far detectors at different times, and the difference between these times, the sonic interval transmit time, is the time taken for the wave to travel through the formation interval. This one transmitter dual detector setup forms the foundation for the currently available sonic tools, which have different numbers of transmitters and detectors in varying configurations and spacings between the transmitters and detectors [207–215].

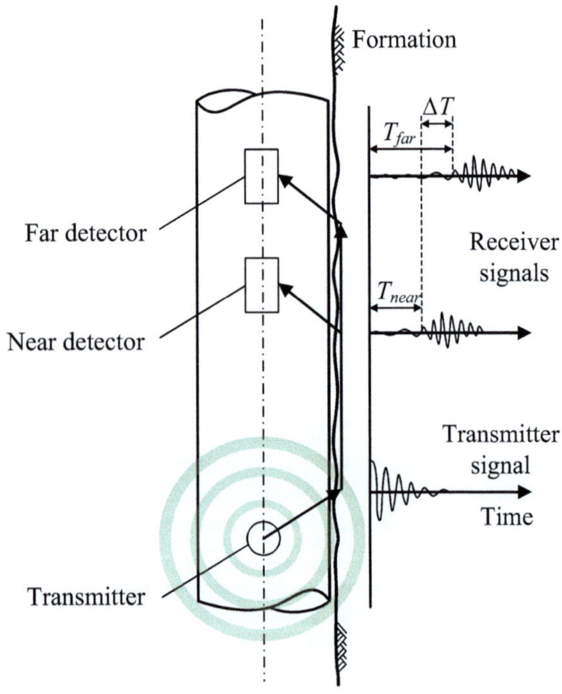

Fig. 2.12 Acoustic logging instrument. A sonic waveform transmitted by the transmitter arrives at the near and far detectors at different times, and the difference between these times, the sonic interval transmit time, is the time taken for the wave to travel through the formation interval. The sonic signal aids in the determination of pore pressure, lithology and geomechanical properties

Three main types of acoustic propagations, compressional, longitudinal or primary waves (P-waves), shear, transverse or secondary waves (S-waves) and Stoneley or Rayleigh surface waves, are analyzed. The P- and S-waves travel through the borehole, where the P-waves are much faster but have smaller amplitudes than S-waves, so are of the most interest. However, S-waves do not travel in fluids since fluids do not have shear strength. The last type of waves to arrive at the detectors are the high amplitude Stoneley/Raleigh waves but they can be easily compensated for in acoustic logs. The time taken for P-waves to arrive at a detector is presented as the travel time per foot and depends mainly on the P-wave propagation time properties of the rock matrix material and the fluids within it. The elastic properties of rock formations can be derived from the S- and P-wave propagation speeds, so acoustic logs are used to determine the dynamic elastic parameters of rock formations and in the identification of formation fractures [1, 3, 6, 8–12]. Solids in general provide a faster propagation medium for P-waves than fluids. As solids become more compact sound propagates faster in them. Therefore, material structure and porosity play a significant role in the travel-time measurement. These relationships can be correlated to porosity and the type and distribution of the pore fluid in the rock formation [216–226].

Acoustic instruments are also used to evaluate the quality of cement jobs after a casing is cemented to isolate a formation section as cementing plays a key role in maintaining the integrity of the well [227–236]. The principle behind the instruments for cement evaluation is the same as explained before and the sound energy transmitted by a transmitter/transmitters travel as waves through casing, cement and rock formations along the axial direction of the instrument and refract back into the drilling fluid in the wellbore to be recorded by detector/detectors. The near detector derives the attenuation factor from the first arrival of the transmitted waveform and the far detector from the full waveform. Acoustic cement evaluation depends on 'knocking on the casing' and listening to the response; if the casing is well bonded the sound will be attenuated and the received amplitude will be low. Because many unknowns are present in the derivation of a cement's compressive strength the amplitude measurement in general can provide only a qualitative estimation of the cement compressive strength, the bond index, which is the ratio of attenuation at any point on the log to the attenuation in the best cemented section. A highly qualitative analysis of the cement to formation interface can be provided by visual analysis of the 5 ft waveform display, known as the variable density log (VDL).

2.2.8 *Imaging*

Wellbore imaging is a virtual visualization tool that provides the driller on the surface high resolution images of rock formations and their surroundings, such as size and orientation of fractures in formations, to optimize the drilling process. The earliest instruments to image formations, named dipmeters, originated about 70 years ago and acquired measurements to determine formation dipping angles

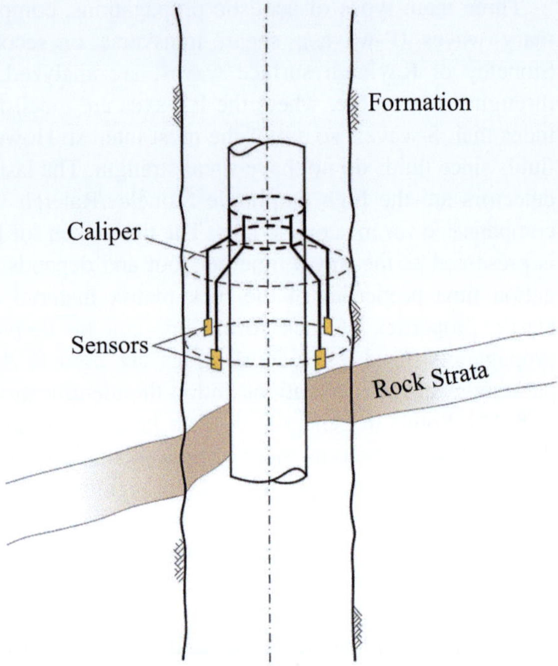

Fig. 2.13 Dipmeter instrument with electrical/mechanical sensor pads that provides structural geometry data of a rock formation. This data is projected to obtain a virtual image of the formation

[237–239]. A dipmeter instrument, as shown in Fig. 2.13, consists of electrical and mechanical sensors on pads that basically acquire wellbore geometry referenced micro-resistivity measurements from the set of pads. The magnitude and direction of formation dips can be derived from these measurements to indicate the structural orientation of rock layers. The early dipmeter tools have progressed to high resolution imaging instruments based on resistivity, acoustic and radioactive measurements that allow a better understanding of geological structures, bedding, pore geometry and geomechanical properties of formations [209, 210, 240–253]. These imaging instruments, initially run only by WL, can now be run by LWD, where the rotation of the instrument while drilling enables azimuthal measurements around the circumference of the wellbore that can be presented as an image of the wellbore. However, while these imaging-while-drilling instruments have a much lower resolution than their WL counterparts they provide valuable information in adjusting the trajectory of a well in real-time. The number of pads and sensors on each pad in both WL and LWD instruments have evolved over the years from one pad-one sensor to multiple pad-multiple sensors configurations [254–260]. The latest WL instrument, also based on microresistivity measurements, have pads with additional flap extensions to increase the circumferential coverage of the wellbore [259, 260]. This design allows measurements with mm resolution of a wellbore. The resolution not only depends on the sensor configuration but also on the electrical contrast with the background, with low resistivities in general being more favourable for obtaining good images. Electrical images are created by converting the microresistivity signal data to pixels and reconstructing them as images.

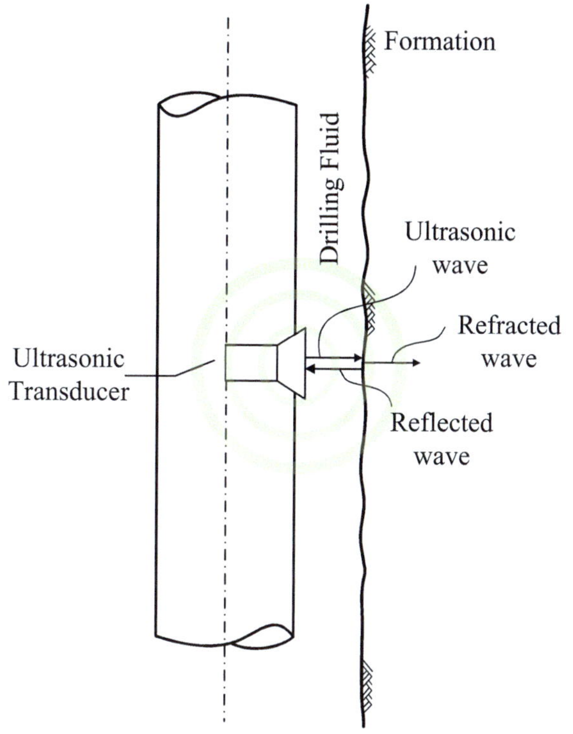

Fig. 2.14 Ultrasonic instrument, where an ultrasonic transducer sends a pulse out and the ratio of transmitted energy to the reflected energy from the rock formation is utilized to obtain information about the formation

Another method is to utilize ultrasonic instruments to obtain images of rock formations [259–265]. High frequency ultrasonic pulses transmitted normal to the wellbore by a transceiver, at frequencies higher than in sonic instruments, in the range of 100–600 kHz, travel through the drilling fluid and contacts the wellbore. At the point of contact most of the energy is reflected and travels back to the transceiver while the rest is refracted through wall of the wellbore into the formation, as shown in Fig. 2.14. Ultrasonic instruments use a technique known as pulse-echo, to map the maximum amplitude of the reflected wave, the echo, as the basis to image the wellbore. While this technique can be used to image the wellbore in the presence of non-conducting drilling fluids it is influenced by the rugosity and shape of the wellbore, the strength of the formation and the density of the mud. Irregularities on the wellbore surface cause the reflected energy to scatter leading to weaker signals, an oval or elliptical shape results in non-normal incidence of the acoustic waves leading to weak echoes or to their complete absence. Additionally, the presence of gas, of large grains and a high concentration of heavy solids in the drilling fluid have strong adverse effects on the quality of ultrasonic logs. In general, ultrasonic imaging has lower resolution compared with modern micro-resistivity tools.

Ultrasonic tools also function as acoustic calipers to measure the diameter of a wellbore, and as cement integrity evaluation instruments. Mechanical calipers, which have pads extending out and pressing against the wellbore, are commonly used to measure the diameter of a wellbore [266–270]. Ultrasonic instruments offer an alternative by correlating the time taken for the transmitted ultrasonic pulse to echo back to the transceiver after contacting the wellbore to the radius of the wellbore. Ultrasonic instruments used for cement evaluation utilize echo amplitudes from interfaces between drilling fluid and inner casing, the inner and outer casing walls, the outer wall of the casing and cement, and a liquid/cement acoustic impedance threshold value to analyze the cement integrity [228, 229, 271–275]. The latest cement evaluation tools uses a combination of the sonic and ultrasonic instruments for more accurate measurements [276, 277].

2.2.9 Fluid Sampling and Testing

Fluid sampling and formation testing instruments take samples of fluid from rock formations and analyze them for the presence of hydrocarbons and also measure the in situ pressure of the formations. Fluid sampling also provides information on composition, gravity and pH of fluids. Early instruments obtained fluid samples by opening a sample chamber by a clock or electric signal from the surface, obtaining the sample and then closing the chamber by a piston [278–281]. The fluids were then brought the fluid samples to the surface for laboratory analysis. Early instruments suffered from sealing problems with the formation when a probe was inserted to draw out fluids, and the drawn fluid samples were usually highly contaminated by the drilling fluids. The very first such tool could obtain only one sample per trip into the well. The rapid advancements made in the 1990s and 2000s has resulted in instruments that are able to obtain multiple uncontaminated samples from the formations in a single trip and perform sophisticated analysis of the sampled fluids even before they are collected, i.e. while the drilling fluid is still in the flowline from which it may be diverted into the sample chamber or discarded into the well. Resistivity, acoustic, densimetric, nuclear magnetic and optical spectroscopy fluid analysis techniques are already available in the formation testing instruments offered by different logging service companies. The downhole analysis results maybe later compared with laboratory analysis results carried in a controlled environment [282–290]. The main advantage of performing downhole analysis as opposed to laboratory analysis is that the fluids can be analyzed at their original downhole pressures downhole and any issues with changes in the fluid composition due to pressure drops while retrieving the samples can be avoided. Additionally, the downhole sensors provide information about fluid contamination making it possible the collection of fluids with a high degree of purity by discarding the contaminated fluids before they get diverted into the sample container. Recently, LWD instruments have also been built for real-time sampling and formation evaluation [291–295].

Figure 2.15 shows a schematic of the general method of sampling and testing. Once a target reservoir interval is chosen for testing with the aid of resistivity/

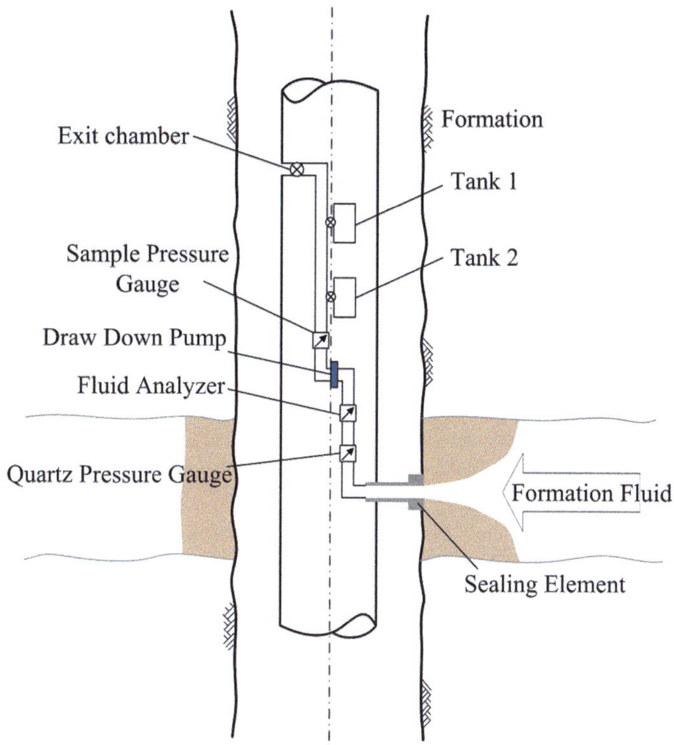

Fig. 2.15 Fluid sampling and testing instrument, which acquires fluid samples from rock formations and use a variety of sensors to test and analyze fluid parameters such as composition, gravity, pH and the presence of hydrocarbons, as well as the pressure of the formation. The tanks are used to store samples for analysis on the surface and the exit chamber is used to dump the fluid samples at any time

density/porosity instruments, the fluid sampling and testing instrument is lowered into the well and anchored at this depth. Ideally the surface of the wellbore at this depth will also have low rugosity and minimum contamination from drilling fluids. A probe containing a piston, surrounded by a seal or packer that presses against the surface of the wellbore and isolates the hydrostatic pressure from the drilling fluid column above, is pushed into the formation and withdrawn by a dual-action piston pump to extract fluids from the formation into the instrument chamber; hydrostatic pressure due to the column of drilling fluid is obtained prior to withdrawing fluids from the reservoir. The pressure sensor allows pressure monitoring of the formation when the fluid is extracted into the instrument, and the pressure behaviour during this drawdown process provides an indication of the reservoir mobility; quartz and strain gauge sensors are used in front and behind the drawdown pump and are normalized to one another for accuracy of measurements. The fluid analyzer may consist of optical spectrometers, resistivity, density, temperature and pressure sensors, to characterize the reservoir fluids and analyze the composition and the contamination levels. The withdrawn fluids are then stored in the sample tanks. The tool operator may choose at any time whether to store or discard the fluid being sampled. This allows the instrument to take as many samples as required and only store uncontaminated samples from the formation in the sample tanks. The fluid sampling and testing instruments currently used in the industry use variations of this basic design and have multiple sample tanks [289, 290].

2.2.10 Seismic

Vertical seismic profiling (VSP) obtains high resolution seismic signals by sampling the acoustic waves travelling between a wellbore and the surface at discrete wellbore depths [296–299]. In comparison, surface seismic has both the source and the receivers on the surface but obtain data as a function of time not depth. VSP has its origins in a method called a check-shot survey, which is used to establish time-depth calibrations and convert surface-recorded seismic images to depth images needed to perform reservoir volumetric calculations. VSP is directly correlated to the geological structure of the rock formations and also provides information on geomechanical, fluid, water saturation and porosity properties of a formation [300–305]. Moreover, it is possible to detect over-pressure zones and drilling risks in advance utilizing VSP and low frequency sonic data; over-pressure zones have formations with pore fluids greater than the normal pressure increasing the risk of dealing with fluid influx into the well and/or drilling fluid losses.

The basic principle of VSP can be explained by referring to Fig. 2.16a, which shows a setup for a WL VSP, with a seismic source at the surface and a number of detectors lowered down into a well. The seismic source is usually a vehicle-mounted seismic vibrator, which can be driven directly to the well site. The detectors, either hydrophones, geophones or accelerometers set about 50–100 ft apart, record the reflected seismic energy originating from the seismic source at the surface. The recorded data is analyzed in the WL truck to obtain a high spatial resolution image of the formations. VSP can be performed with a number of different source-receiver configurations, such as zero-offset, offset, walk-away, cross-well, intra-well and inverted, to obtain 2D and 3D VSP of complex rock formations and their boundaries [300, 303]. Seismic-while-drilling obtains real-time data while drilling by having a source at the surface and an array of detectors at the BHA [300, 303, 306–310]. On the other hand, drill-bit listening VSP, also known as drill bit seismic-while-drilling, is a VSP technique that uses the drill bit as the acoustic source and detectors at the surface, as shown in Fig. 2.16b. The noise created by the drill bit penetrating formations and crushing rocks is propagated from downhole to the detectors placed at the surface [300, 311–315]. The drill bit wave field while drilling creates an image of the formations being drilled through. Drill-bit seismic-while-drilling VSP is influenced by the type of drill bit and the speed used to penetrate through formations.

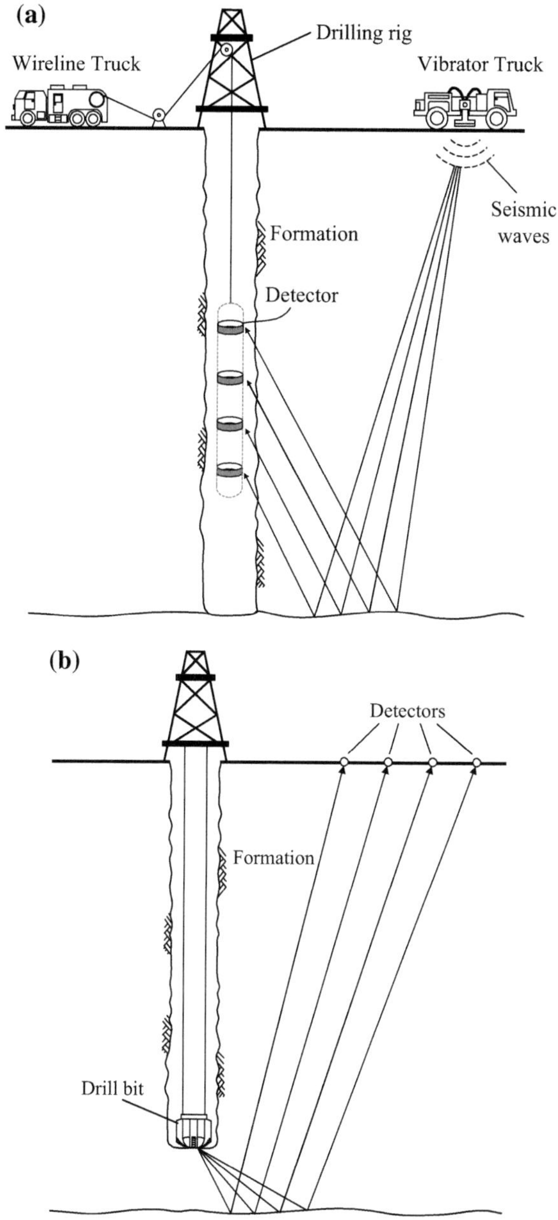

Fig. 2.16 High spatial resolution images obtained by vertical seismic profiling (VSP) and seismic-while-drilling. **a** In vertical seismic profiling seismic waves are transmitted into the formation utilizing a vehicle-mounted vibrator and the reflected waves are recorded by a series of detectors lowered into the well by wireline. **b** In drill bit seismic-while-drilling, the noise created by the drill bit drilling through formations is recorded by detectors at the surface

References

1. D.V. Ellis, *Well Logging for Earth Scientists* (Springer, 2007)
2. L.M. Etnyre, *Finding Oil and Gas from Well Logs* (Springer, 1989)
3. O. Serra, *Fundamentals of Well-Log Interpretation—The Acquisition of Logging Data* (Elsevier, 1984)
4. B. Pike, R. Duey, *Hart's E&P*, pp. 52–55 (2002). Available online at https://www.spwla.org/SPWLA/About/History/Logging_history_rich_with_innovation__Hart_s_E_P_Magazine_.aspx
5. D.E. Johnson, K.E. Pile, *Well Logging in Nontechnical Language* (PennWell Books, 2002)
6. G.B. Asquith, D. Krygowski, C.R. Gibson, *Basic Well Log Analysis* (American Association of Petroleum Geologists, Tulsa, 2004)
7. M.A. Andersen, Oilfield Rev. **23**, 58 (2011)
8. R.M. Bateman, *Openhole Log Analysis and Formation Evaluation* (Society of Petroleum Engineers, 2012)
9. Z. Bassiouni, *Theory, Measurement, and Interpretation of Well Logs* (Society of Petroleum Engineers, 1994)
10. J. Evenick, *Introduction to Well Logs and Subsurface Maps* (PennWell Books, 2008)
11. D.W. HIlchie, *Wireline: a History of the Well Logging and Perforating Business in the Oil Fields* (DW Hilchie, 1990)
12. Schlumberger Educational Services, *Log Interpretation Principles/Applications* (1989). Available online at https://www.slb.com/resources/publications/books/lipa.aspx
13. M. Billingham, A.M. El-Toukhy, M.K. Hashem, M. Hassaan, M. Lorente, T. Sheiretov, M. Loth, Oilfield Rev. **23**, 18 (2011)
14. S. Prensky, Petrophysics **43** (2002)
15. D. Hoyle, V. Tourillon, P. Webb, H. Soliman, in *SPE/IADC Drilling Conference* (1996)
16. M. Liyanage, J. Tulett, T. Endo, A. Saito, H. Yamamoto, R. Imai, in *SPWLA 22nd Formation Evaluation Symposium of Japan* (2016)
17. F. Jahn, M. Cook, M. Graham, *Hydrocarbon Exploration and Production* (Elsevier, 2008)
18. D. Ghiselin, *Offshore Magazine* (2012). Available online at http://www.offshore-mag.com/articles/print/volume-72/issue-8/drilling-and-completion/logging-while-drilling-technology-continues-to-reduce-drilling-risk.html
19. G. Weller, R. Griffiths, C. Stoller, F. Allioli, in *SPWLA 46th Annual Logging Symposium* (2005)
20. S. Prensky, World Oil **69** (2006). Available online at https://www.slb.com/~/media/files/drilling/industry_articles/wo_mwd_formation_evaluation.pdf
21. D. Allen, D. Bergt, D. Best, B. Clark, I. Falconer, J.M. Hache, C. Kienitz, M. Lesage, J. Rasmus, C. Roulet, P. Wraight, Oilfield Rev. **1**, 4 (1989)
22. R.R. Hansen, J. White, in *SPE/IADC Drilling Conference* (1991)
23. S. Bonner, B. Clark, J. Holenka, B. Voisin, J. Dusang, R. Hansen, J. White, T. Walsgrove, Oilfield Rev. **4**, 4 (1992)
24. S. Bargach, I. Falconer, C. Maeso, J. Rasmus, T. Bornemann, R. Plumb, D. Codazzi, K. Hodenfield, G. Ford, J. Hartner, B. Grether, Oilfield Rev. **12**, 58 (2000)
25. P. Hansen, F. Shray, in *SPWLA 37th Annual Logging Symposium* (1996)
26. C.E. Jackson, D.R. Heysse, in *SPE Annual Technical Conference and Exhibition* (1994)
27. P. Boonen, in AADE
28. M.A. Frenkel, I. Nuic, M.J. Walker, F.L. Wolters, in *SPE Asia Pacific Oil and Gas Conference and Exhibition* (2004)
29. V.A. Briggs, A comparison of logging while drilling (LWD) and wireline acoustic measurements, Ph.D. thesis, 2006. Available online at https://dspace.mit.edu/handle/1721.1/37981
30. A. Mendoza, D. Ellis, J.C. Rasmus, in *International Oil Conference and Exhibition in Mexico* (2006)

31. C.A. de Andr, A.M. Vieira da Cunha, P. Boonen, B. Valant-Spaight, S. LeFors, W. Schultz, Petrophysics **46** (2005)
32. D.J. Seifert, R. Akkurt, S. Al-Dossary, R.M. Shokeir, H.V. Ersoz, in *SPE Middle East Oil and Gas Show and Conference* (2007)
33. E.M. Tollefsen, A.B. Kramer, L.S Grant, in *International Oil Conference and Exhibition in Mexico* (2007)
34. J.Market, J. Quirein, J.L. Pitcher, D. Hinz, D. Buller, C. Al-dammad, D.R. Spain, T. Odumosu, in *SPE Annual Technical Conference and Exhibition* (2010)
35. M. Rabinovich, F. Le, J.J. Lofts, S. Martakov, Petrophysics **53**, 86 (2012)
36. A.M. Serry, S.A. Budebes, H. Aboujmeih, A. Aki, M. Bittar, in *SPWLA 55th Annual Logging Symposium*, Society of Petrophysicists and Well-Log Analysts (2014)
37. G.E. Archie, Trans. Am. Inst. Min. Eng. **146**, 54 (1942)
38. S. Bonner, M. Fredette, J. Lovell, B. Montaron, R. Rosthal, J. Tabanou, P. Wu, B. Clark, R. Mills, R. Williams, Oilfield Rev. **8**, 19 (1996)
39. I.L. Roberts, in *Formation Evaluation Symposium* (1955)
40. H.G. Doll, J. Petrol. Technol. **5**, 33 (1953)
41. W.W. Whitman, Log Anal. **36** (1995)
42. C.R. Liu, *Theory of Electromagnetic Well Logging* (Elsevier, 2017)
43. H. Doll, J. Petrol. Technol. **3**, 305 (1951)
44. A.A. Fitch, *Developments in Geophysical Exploration Methods* (Springer, 1981)
45. K.M Strack, O.N. Fanini, A.G. Mezzatesta, L.A. Tabarovsky, Log Anal. **39** (1998)
46. H. Guyod, J. Petrol. Technol. **16**, 211 (1964)
47. H. Guyod, Log Anal. **7**, 27 (1966)
48. J. Suau, P. Grimaldi, A. Poupon, P. Souhaite, in *Fall Meeting of the Society of Petroleum Engineers of AIME* (1972)
49. W.H. Fertl, G.W. Hammack, Log Anal. **13** (1972)
50. R. Chemali, S. Gianzero, S.M. Su, in *SPWLA 29th Annual Logging Symposium*, Society of Petrophysicists and Well-Log Analysts (1988)
51. H.M. Wang, L.C. Shen, G.J. Zhang, in *SPWLA 39th Annual Logging Symposium*, Society of Petrophysicists and Well-Log Analysts (1998)
52. L.C. Shen, H. Wang, Petrophysics **42** (2001)
53. W. Yang, C. Torres-Verdín, R. Akkurt, S. Al-Dossari, A. Al-Towijri, H. Ersoz, in *Society of Petrophysicists and Well Log Analysts, Expanded Abstract* (2007)
54. L.A. Lander, A. Silva, C. Simon, in *SPE Latin American and Caribbean Petroleum Engineering Conference* (2015)
55. J. Li, L.C. Shen, Log Anal. **33** (1992)
56. D.H. Davies, O. Faivre, M.T. Gounot, B. Seeman, J.C. Trouiller, D. Benimeli, A.E. Ferreira, D.J. Pittman, J.W. Smits, M. Randrianavony, B.I. Anderson, SPE Form. Eval. **9**, 165 (1994)
57. J.W. Smits, D. Benimeli, I. Dubourg, O. Faivre, D. Hoyle, V. Tourillon, J.C. Trouiller, B.I. Anderson, in *SPE Annual Technical Conference and Exhibition* (1995)
58. M.T. Ribeiro, I. Hendrawan, S. Karrani, R.J. Griffiths, R.M. Greiss, S. Chakravorty, M. Bazara, B. Sirum, in *SPE/IADC Middle East Drilling Technology Conference and Exhibition* (2005)
59. M. Allouche, S. Chow, I. Dubourg, L. Ortenzi, R. van Os, in *SPE EUROPEC/EAGE Annual Conference and Exhibition* (2010)
60. P. Sinclair, A. Li, A. Ottemen, A. Wennekamp, in *IADC/SPE Asia Pacific Drilling Technology Conference and Exhibition* (2012)
61. J.W. Smits, I. Dubourg, M.G. Luling, G.N. Minerbo, J.M.V.A. Koelman, L.J.B. Hoffman, A.T.Lomas, R. Oosten, M.J. Schiet, R.N. Dennis, in *SPE Annual Technical Conference and Exhibition* (1998)
62. Y.H. Chen, W.C. Chew, G.J. Zhang, Log Anal. **39** (1998)
63. T. Wang, A. Mezzatesta, L. Xiao, in *1999 SEG Annual Meeting of the Society of Exploration Geophysicists* (1999)

64. E. Legendre, I. Dubourg, J. Doduy, J.W. Smits, O. Faivre, R. Griffiths, in *SPWLA 40th Annual Logging Symposium* (1999)
65. H. Maurer, Y. Antonov, B. Corley, R. Khokhar, M. Rabinovich, Z. Zhou, in *SPWLA 50th Annual Logging Symposium* (2009)
66. J.M. Donadille, I. Ilyin, M.G. Lüling, T. Meszaros, R. Reischman, A. Repnev, V. Yukhlin, N. Khan, M. Jones, in *Unconventional Resources Technology Conference* (2017)
67. H.G. Doll, J. Petrol. Technol. **5**, 17 (1953)
68. P.D. Jackson, Geophys. Prospect. **29**, 601 (1981)
69. A.G. Barminsky, R.A. Kucherov, N.I. Tanzurenko, Y.Y. Tolmachev, Log Anal. **32** (1991)
70. K.M. Strack, M.A. Frenkel, J.M.V.A. Koelman, A.G. Mezzatesta, S. Talabani, in *8th Abu Dhabi International Petroleum Exhibition and Conference* (1998)
71. S. Bhattacharya, J.H. Doveton, T.R. Carr, W.R. Guy, P.M. Gerlach, AAPG Bull. **89**, 1257 (2005)
72. B. Corley, R. Khokhar, H.M. Maurer, J.Z. Li, D. Xiao, S. Yang, in *SPE Oil and Gas India Conference and Exhibition* (2012)
73. H.G. Doll, Petrol. Trans. Am. Inst. Min. Eng. Tech. Pap. **2641** (1949)
74. J.H. Moran, K.S. Kunz, Geophysics **27**, 829 (1962)
75. B. Kriegshauser, O. Fanini, S. Forgang, G. Itskovich, M. Rabinovich, L. Tabarovsky, L. Yu, M. Epov, P. Gupta, in *SPWLA 41st Annual Logging Symposium*, Society of Petrophysicists and Well-Log Analysts (2000)
76. M. Rabinovich, B. Corley, D. Georgi, L. Taborovsk, M. Gonfalini, T. Rocque, M. Epov, in *48th Annual Logging Symposium* (2007)
77. B. Anderson, T. Barber, R. Leveridge, R. Bastia, K.R. Saxena, A.K. Tyagi, J.B. Clavaud, B. Coffin, M. Das, R. Hayden, T. Klimentos, Analyst **23**, 20 (1982)
78. J. Xiao, J.C. Buchanan, M.S. Bittar, E. Davies, L. Sanmartin, G. Hu, S. Zannoni, M. Morys, W. Liu, in *SPE Annual Technical Conference and Exhibition* (2006)
79. J. Hou, L.S. Martin, D. Wu, F.T. Celepcikay, D. Torres, in *SPWLA 53rd Annual Logging Symposium*, Society of Petrophysicists and Well-Log Analysts (2012)
80. M.G. Persova, Y.G. Soloveichik, D.V. Vagin, Y.I. Koshkina, in *Proceedings of the Actual Problems of Electronics Instrument Engineering (APEIE)*, vol. 2, p. 305 (2016)
81. S.C. Gianzero, IEEE Trans. Geosci. Electron. **16**, 332 (1978)
82. J.H. Moran, Log Anal. **23** (1982)
83. X. Tian, C. Liu, L.C. Shen, Radio Sci. **33**, 949 (1998)
84. A.A. Gorbatenko, K.V. Sukhorukova, Russ. Geol. Geophys. **57**, 1111 (2016)
85. H. Wang, H. Tao, J. Yao, Y. Zhang, IEEE Geosci. Remote Sens. **50**, 3383 (2012)
86. F. Khoshbakht, M.R. Rasaie, A. Shekarifard, J. Petrol. Sci. Eng. **145**, 357 (2016)
87. Y. Chen, Z. Cheng, J. Wu, Y. Mao, S. Luo, F. Xiao, Z. Yang, H. Gao, in *Progress in Electromagnetic Research Symposium (PIERS)*, pp. 3359–3363 (2016)
88. L. Zhong, J. Li, A. Bhardwaj, L.C. Shen, R.C. Liu, IEEE Geosci. Remote Sens. **46**, 1148 (2008)
89. W.H. Meyer, L.W. Thompson, M.M. Wisler, J.Q. Wu, in *SPWLA 35th Annual Logging Symposium* (1994)
90. T. Schroeder, in *SPWLA 42nd Annual Logging Symposium* (2001)
91. G. Hu, M.S. Bittar, J. Hou, in *SPE Annual Technical Conference and Exhibition* (2006)
92. S.M. Haugland, in *SPWLA 42nd Annual Logging Symposium* (2001)
93. B. Clark, D.F. Allen, D.L. Best, S.D. Bonner, J. Jundt, M.G. Luling, M.O. Ross, SPE Form. Eval. **5**, 263 (1990)
94. W.H. Meyer, Petrophysics **41** (2000)
95. M.S. Bittar, G. Hu, J. Xiao, in *SPE Annual Technical Conference and Exhibition* (2003)
96. M. Haugland, in *SPE Annual Technical Conference and Exhibition* (2005)
97. M.G. Prammer, M. Morys, S. Knizhnik, C.J. Conrad, W.E. Hendricks, M.S. Bittar, G. Hu, F. Hveding, K. Kenny, R.M. Shokeir, D.J. Seifert, in *48th Annual Logging Symposium* (2007)
98. J.L. Pitcher, M.S. Bittar, D. Hinz, C. Knutson, R. Cook, in *SPE EUROPEC/EAGE Annual Conference and Exhibition* (2011)

99. J. Li, M.K. Dennis, J. Rasmus, D. Kennedy, T. Wang, in *SPWLA 53rd Annual Logging Symposium*, Society of Petrophysicists and Well-Log Analysts (2012)
100. Z. Qin, H. Pan, A. Wu, H. Yang, T. Hu, M. Hou, Z. Wang, J. Geophys. Eng. **14**, 1233 (2017)
101. M.I. Epov, V.N. Glinskikh, V.N. Eremin, M.N. Nikitenko, A.N. Petrov, I.V. Mikhaylov, in *SPE Russian Petroleum Technology Conference* (2017)
102. R.P. Wharton, R.N. Rau, D.L. Best, in *SPE Annual Technical Conference and Exhibition* (1980)
103. R.P. Wharton, G.A. Hazen, R.N. Rau, D.L. Best, in *SPE Rocky Mountain Regional Meeting* (1980)
104. L.C. Shen, M.J. Manning, J.M. Price, in *SPWLA 25th Annual Logging Symposium* (1984)
105. M. Blenkinsop, P. Baker, C. Clavier, W. Kenyon, S. des Igneris, in *SPWLA 27th Annual Logging Symposium* (1986)
106. T.J. Calvert L.E. Wells, in *SPE California Regional Meeting* (1977)
107. W.C. Chew, IEEE Trans. Geosci. Remote Sens. **26**, 382 (1988)
108. R. Rau, R. Davies, M. Finke, M. Manning, in *SPWLA 32nd Annual logging symposium* (1991)
109. L.C. Shen, W.C. Savre, J.M. Price, K. Athavale, Geophysics **50**, 692 (1985)
110. D.P. Schmitt, A. Al-Harbi, P. Saldungaray, R. Akkurt, T. Zhang, in *SPE/DGS Saudi Arabia Section Technical Symposium and Exhibition* (2011)
111. M.T. Frey, L. Argentier, M. Ross, V. Varveropoulos, in *SPE Annual Technical Conference and Exhibition* (2006)
112. J. Mude, S. Arora, T. McDonald, J. Edwards, in *SPWLA 51st Annual Logging Symposium* (2010)
113. N.V. Seleznev, R.L. Kleinberg, M.M. Herron, M. Machlus, A.E. Pomerantz, S.L. Reeder, A. K. Burnham, R.L. Day, P.C. Allix, in *SPWLA 52nd Annual Logging Symposium* (2011)
114. H.G. Doll, Trans. AIME **179**, 146 (1949)
115. E.A. Koerperich, J. Petrol. Technol. **22**, 1437 (1970)
116. J.K. Hallenburg, in *SPWLA 12th Annual Logging Symposium* (1971)
117. G.G. Tenchov, J. Petrol. Sci. Eng. **7**, 309 (1992)
118. J.R. Tabanou, R. Glowinski, G.F. Rouault, in *SPWLA 28th Annual Logging Symposium* (1987)
119. J. Bautista-Anguiano, C. Torres-Verdín, in *SPWLA 56th Annual Logging Symposium* (2015)
120. L.G. Howell, A. Frosch, Geophysics **4**, 106 (1939)
121. J.K. Hallenburg, Log Anal. **14** (1973)
122. D.V. Ellis, J.S. Schweitzer, J.J. Ullo, Annu. Rev. Nucl. Part. Sci. **37**, 213 (1987)
123. A.J. Zak, J.E. Smith, in *Permian Basin Oil Recovery Conference* (1959)
124. J.S. Wahl, Geophysics **48**, 1536 (1983)
125. Y. Xiao, S. Xiang, Z. Zhao, Z. Qian, Procedia Eng. **7**, 223 (2010)
126. W. Bertozzi, D.V. Ellis, J.S. Wahl, Geophysics **46**, 1439 (1981)
127. D.V. Ellis, IEEE Trans. Nucl. Sci. **35**, 806 (1988)
128. S.A. Kerr, P.F. Worthington, IEEE Trans. Nucl. Sci. **35**, 794 (1988)
129. J.A. Quirein, J.S. Gardner, J.T. Watson, in *SPE Annual Technical Conference and Exhibition* (1982)
130. G.L. Moake, W.E. Schultz, in *Middle East Oil Show* (1989)
131. C.J. Mwenifumbo, B.E. Elliott, K.A. Pflug, M. Salisbury, Petrophysics **46** (2005)
132. J.A. Sanchez-Ramirez, C. Torres-Verdín, G.L. Wang, A. Mendoza, D. Wolf, Z. Liu, G. Schell, in *SPWLA 50th Annual Logging Symposium* (2009)
133. M. Luycx, C. Torres-Verdín, in *SPE Annual Technical Conference and Exhibition* (2017)
134. S.G. Carrilero, D. Hinz, G. Moake, T. Parker, D. Carson, in *International Petroleum Technology Conference* (2016)
135. K. Lehmann, J. Appl. Geophys. **70**, 17 (2010)
136. H.R. Brannon Jr., J.S. Osoba, Trans. Soc. Petrol. Eng. AIME **207**, 30 (1956)

137. R.L. Caldwell, W.F. Baldwin, J.D. Bargainer, J.E. Berry, G.N. Salaita, R.W. Sloan, Geophysics **28**, 617 (1963)
138. L.L. Gadeken, H.D. Smith, D.M. Arnold, C.A. Robbins, IEEE Trans. Nucl. Sci. **35**, 822 (1988)
139. G.L. Mathis, C.W. Tittle, D.R. Rutledge, R. Mayer Jr., W.E. Ferguson, in *SPWLA 25th Annual Logging Symposium* (1984)
140. W.H. Fertl, Log Anal. **20** (1979)
141. J. Klaja, L. Dudek, Nafta-Gaz **72**, 3 (2016). https://doi.org/10.18668/NG2016.01.01
142. L. Xu, C. Huiszoon, J. Wang, B. Adolph, J. Yi, D. Cavin, G. Laughlin, E. Tollefsen, S. Jacobsen, M. Boyce, in *SPWLA 56th Annual Logging Symposium* (2015)
143. C. Maranuk, M. Mickael, P. Zimmermann, in *SPWLA 54th Annual Logging Symposium* (2013)
144. A. Kepic, A. Podolska, I. Hooshyari Far, A. Greenwood, in *Proceedings of the Near-Surface Asia Pacific Conference*, pp. 220–223 (2015)
145. B. Pontecorvo, Oil Gas J. **40**, 32 (1941)
146. L.S. Allen, C.W. Tittle, W.R. Mills, R.L. Caldwell, Geophysics **32**, 60 (1967)
147. B.G. Hubner, W.D. Bruck, J.H. Dudek, in *IEEE Nuclear Science Symposium Conference Record* (1990)
148. J. Holenka, D. Best, M. Evans, P. Kurkoski, W. Sloan, in *SPWLA 36th Annual Logging Symposium* (1995)
149. D.V. Ellis, C.R. Case, J.M. Chiaramonte, Petrophysics **44** (2003)
150. A. Nikitin, S. Bliven, in *Proceedings of the Nuclear Science Symposium Conference Record*, pp. 1214–1219 (2010)
151. W. Wu, M. Tong, L. Xiao, J. Wang, J. Petrol. Sci. Eng. **108**, 10 (2013)
152. D. Hilchie, W.R. Mills, C.L. Dennis, W.W. Givens, in *SPWLA 9th Annual Logging Symposium* (1968)
153. R.C. Hertzog, R.E. Plasek, IEEE Trans. Nucl. Sci. **26**, 1558 (1979)
154. C. Flaum, G. Pirie, in *SPWLA 22nd Annual Logging Symposium* (1981)
155. W.A. Gilchrist Jr., J.A. Quirein, Y.L. Boutemy, J.R. Tabanou, in *SPWLA 23rd Annual Logging Symposium* (1982)
156. J. Galford, J. Truax, A. Hrametz, C. Haramboure, in *SPWLA 50th Annual Logging Symposium* (2009)
157. J.E. Galford, J.A. Quirein, S. Shannon, J.A. Truax, J. Witkowsky, in *SPE Annual Technical Conference and Exhibition* (2009)
158. D. Rose, T. Zhou, S. Beekman, T. Quinlan, M. Delgadillo, G. Gonzalez, S. Fricke, J. Thornton, D. Clinton, F. Gicquel, I. Shestakova, in *SPWLA 56th Annual Logging Symposium* (2015)
159. R. Odom, T. Dahlin, D. Dolliver, E. Elizondo, E. Seeds, J. Starnes, R. Wilson, in *49th Annual Logging Symposium* (2008)
160. F. Zhang, L. Tian, J. Liu, Q. Zhang, X. Wang, Q. Chen, in *SEG International Exposition and Annual Meeting* (2016)
161. H. Yu, J. Sun, J. Wang, R.P. Gardner, Appl. Radiat. Isot. **69**, 1313 (2011)
162. A.A. Arbuzov, A.P. Alekhin, V.V. Bochkarev, R.N. Minakhmetova, D.V. Chukhutin, A.N. Zakirov, in *SPE Russian Oil and Gas Exploration and Production Technical Conference and Exhibition* (2012)
163. R.J.S. Brown, B.W. Gamson, J. Petrol. Technol. **210** (1960)
164. R.J.S. Brown, Concepts Magn. Reson. **13**, 344 (2001)
165. R.L. Kleinberg, J.A. Jackson, Concepts Magn. Reson. **13**, 340 (2001)
166. R.K. Cooper, J.A. Jackson, J. Magn. Reson. **41**, 400 (1980)
167. L.J. Burnett, J.A. Jackson, J. Magn. Reson. **41**, 406 (1980)
168. J.A. Jackson, L.J. Burnett, F. Harmon, J. Magn. Reson. **41**, 411 (1980)
169. M. Lipsicas, J.R. Banavar, J. Willemsen, Appl. Phys. Lett. **48**, 1554 (1986)
170. R. Kleinberg, W. Kenyon, P. Mitra, J. Magn. Reson. Ser. A **108**, 206 (1994)

171. R.W. Mair, G. Wong, D. Hoffmann, M.D. Hürlimann, S. Patz, L.M. Schwartz, R.L. Walsworth, Phys. Rev. Lett. **83**, 3324 (1999)
172. M. Hurlimann, K.G. Helmer, L. Latour, C.H. Sotak, J. Magn. Reson. Ser. A **111**, 169 (1994)
173. M. Hürlimann, M. Flaum, L. Venkataramanan, C. Flaum, R. Freedman, G. Hirasaki, Magn. Reson. Imaging **21**, 305 (2003)
174. M.D. Hürlimann, L. Venkataramanan, J. Magn. Reson. **157**, 31 (2002)
175. Y.-Q. Song, L. Venkataramanan, R. Kausik, N. Heaton, in *Diffusion NMR of Confined Systems: Fluid Transport in Porous Solids and Heterogeneous Materials*, chap. 4, ed. by R. Valiullin (The Royal Society of Chemistry, 2016)
176. F. Balibanu, K. Hailu, R. Eymael, D. Demco, B. Blumich, J. Magn. Reson. **145**, 246 (2000)
177. F. Casanova, J. Magn. Reson. **163**, 38 (2003)
178. M. Todica, R. Fechete, B. Blümich, J Magn. Reson. **164**, 220 (2003)
179. R. Akkurt, H.N. Bachman, C.C. Minh, C. Flaum, J. LaVigne, R. Leveridge, R. Carmona, S. Crary, E. Decoster, N. Heaton, M.D. Hurlimann, W.J. Looyestijn, D. Mardon, J. White, Oilfield Rev. **20**, 4 (2009)
180. R.L. Kleinberg, A. Sezginer, D.D. Griffin, M. Fukuhara, Novel NMR apparatus for investigating an external sample. J. Magn. Reson. **97**, 466 (1969, 1992)
181. M.N. Miller, Z. Paltiel, M.E. Gillen, J. Granot, J.C. Bouton, in *SPE Annual Technical Conference and Exhibition* (1990)
182. M. Appel, Petrophysics **45** (2004)
183. A. Timur, Log Anal. **10** (1969)
184. U. Ahmed, S.F. Crary, G.R. Coates, J. Petrol. Technol. **43**, 578 (1991)
185. G. Hursan, A. Silva, M.L. Zeglache, in *SPE Annual Technical Conference and Exhibition* (2016)
186. R. Akkurt, A.F. Marsala, D. Seifert, A. Al-Harbi, C. Buenrostro, T. Kruspe, H.F. Thern, G. Kurz, M. Blanz, A. Kroken, in *SPE Saudi Arabia Section Technical Symposium* (2009)
187. M.G. Prammer, E. Drack, G. Goodman, P. Masak, S. Menger, M. Morys, S. Zannoni, B. Suddarth, J. Dudley, SPE Reservoir Eval. Eng. **4**, 270 (2001)
188. R.J. Alvarado, A. Damgaard, P. Hansen, M. Raven, R. Heidler, R. Hoshun, J. Kovats, C. Morriss, D. Rose, W. Wendt, Oilfield Rev. **40** (2003)
189. R. Heidler, C. Morriss, R. Hoshun, in *SPWLA 44th Annual Logging Symposium* (2003)
190. M. Blanz, T. Kruspe, H.F. Thern, G.A. Kurz, Diffus. Fundam. **14**, 1 (2010). Available online at http://ul.qucosa.de/fileadmin/data/qucosa/documents/18630/diff_fund_14%282010%292.pdf
191. A. Kurup, A. Valori, H. N. Bachman, J.-P. Korb, M. Hürlimann, L. Zielinski, in *SPE Saudi Arabia Section Technical Symposium and Exhibition* (2013)
192. M.D. Hürlimann, D.E. Freed, L.J. Zielinski, Y.Q. Song, G. Leu, C. Straley, A. Boyd, C.C. Minh, Petrophysics **50**, 116 (2009)
193. R. Kausik, M.D. Hürlimann, J. Magn. Reson. **270**, 12 (2016)
194. R. Akkurt, D. Kersey, Zainalabedin. Petrophysics **50**, 498 (2009)
195. D.J. Seifert, R. Akkurt, S. Al-Dossary, R. Shokeir, H. Ersoz, in *SPE Middle East Oil and Gas Show and Conference* (2007)
196. R. Freedman, J. Petrol. Technol. **58**, 60 (2006)
197. S. Menger, P. Sands, Petrophysics **44**, 190 (2003)
198. M.R.J. Wyllie, A.R. Gregory, L.W. Gardner, Geophysics **21**, 41 (1956)
199. M.R.J. Wyllie, A.R. Gregory, G.H.F. Gardner, Geophysics **23**, 459 (1958)
200. G.R. Pickett, J. Petrol. Technol. **15**, 659 (1963)
201. K. Hsu, T.L. Marzetta, IEEE Trans. Acoust. Speech Signal Process. **37**, 265 (1989)
202. D.P. Schmitt, D.M. Williams, J. Zemanek, Log Anal. **32** (1991)
203. C.H.Cheng, F.L. Paillet, W.D. Pennington, Log Anal. **33** (1992)
204. D. Close, D. Cho, F. Horn, H. Edmundson, CSEG Recorder **34**, 34 (2009)
205. Y. Zhou, X. Wang, Y. Dai, J. Acoust. Soc. Am. **138**, 1827 (2015)
206. X. Song, Y. Zhao, J. Dykstra, IEEE Trans. Ind. Electron. **64**, 3423 (2017)
207. X.M. Tang, V. Dubinsky, A. Bolshakov, D. Patterson, Petrophysics **44**, 79 (2003)

208. C. Kessler, G. Varsamis, J. Blanch, A. Cheng, in *SPWLA 45th Annual Logging Symposium* (2004)
209. V. Pistre, T. Kinoshita, T. Endo, K. Schilling, J. Pabon, in *SPWLA 46th Annual Logging Symposium* (2005)
210. J.B. Haldorsen, D.L. Johnson, T. Plona, B. Sinha, H.P. Valero, K. Winkler, Oilfield Rev. **18**, 34 (2006)
211. T. Kinoshita, A. Dumont, H. Hori, N. Sakiyama, J. Morley, F. Garcia-Osuna, in *SEG Technical Program Expanded Abstracts*, pp. 513–517 (2010)
212. J.M. Degrange, A. Hawthorn, H. Nakajima, T. Fujihara, M. Mochida, in *IADC/SPE Drilling Conference and Exhibition* (2010)
213. A. Chikenji, D. Scheibner, H. Yamamoto, P. Sah, H.P. Valero, A. Shirota, M. Tohda, M. Ito, Q. Yang, Y. Sakata, in *SPWLA 52nd Annual Logging Symposium* (2011)
214. J. Alford, M. Blyth, E. Tollefsen, J. Crowe, J. Loreto, S. Mohammed, V. Pistre, A. Rodriguez-Herrera, Oilfield Rev. **24**, 4 (2012)
215. J.C. Auchere, Y. Hatori, A. Oshima, H. Yamamoto, H. Hori, T. Fujihara, H. Nakajima, H. P. Valero, in *19th Formation Evaluation Symposium of Japan* (2013)
216. J.O. Guy, B. Corley, W.H. Fertl, E. Frost, in *SPWLA 28th Annual Logging Symposium* (1987)
217. J.W. Minear, Southeast Asia Show (1984)
218. J.W. Minear, D.R. Heysse, P.M. Boonen, *SPE Annual Technical Conference and Exhibition* (1996)
219. D.R. Heysse, C. Robbins, J. Minear, in *SPWLA 37th Annual Logging Symposium* (1996)
220. K. Hsu, M. Hashem, C.L. Bean, R. Plumb, G.N. Minerbo, in *SPWLA 38th Annual Logging Symposium* (1997)
221. A. Brie, T. Endo, D. Hoyle, D. Codazzi, C. Esmersoy, K. Hsu, S. Denoo, M.C. Mueller, T. Plona, R. Shenoy, B. Sinha, Oilfield Rev. **10**, 40 (1998)
222. J.A. Donald, T.R. Bratton, in *SPWLA 47th Annual Logging Symposium* (2006)
223. E. Dyshlyuk, A.V. Parshin, M.G. Charara, A. Nikitin, B. Plyushchenkov, in *SPE Annual Technical Conference and Exhibition* (2011)
224. J. Cai, L. Zhen, S.Y. Han, J. YiMing, in *Offshore Technology Conference*, May 2014
225. R. Wang, C. Torres-Verdín, S. Huang, W. Herrera, in *SPWLA 56th Annual Logging Symposium* (2015)
226. B. Sun, R. Wang, W. Li, B.S. Jonathan, B. Langley, G. Wilson, A. Cheng, in *Offshore Technology Conference Asia* (2016)
227. G.H. Pardue, R.L. Morris, J. Petrol. Technol. **15**, 545 (1963)
228. R.M. Havira, in *SPWLA 23rd Annual Logging Symposium* (1982)
229. A.J. Hayrnan, R. Hutin, P.V. Wright, in *SPWLA 32nd Annual Logging Symposium* (1991)
230. S.I. Watanabe, W. Izuhara, V. Pistre, H. Yamamoto, in *20th Formation Evaluation Symposium of Japan* (2014)
231. T. Kinoshita, W. Izuhara, H.P. Valero, M. Blyth, in *SPE Annual Technical Conference and Exhibition* (2013)
232. M. Blyth, D. Hupp, I. Whyte, T. Kinoshita, in *SPE/IADC Drilling Conference* (2013)
233. P.X. Li, X.M. Zhang, W.J. Lin, in *Symposium on Piezoelectricity, Acoustic Waves, and Device Applications*, pp. 130–133 (2014)
234. G.A. Fuller, N. Gupta, M. Blyth, A. Aparicio, in *SPE/IADC Drilling Conference and Exhibition* (2015)
235. J. Tang, C. Zhang, B. Zhang, F. Shi, Cement bond quality evaluation based on acoustic variable density logging. Petrol. Explor. Dev. **43**, 514–521 (2016)
236. H. Wang, G. Tao, X. Shang, J. Acoust. Soc. Am. **139**, 2407 (2016)
237. H.G. Doll, Petrol. Technol. **6**, 1 (1943)
238. A.P. Claudet, *Drilling and Production Practice* (American Petroleum Institute, 1950)
239. L.A. Alluad, J. Ringot, Log Anal. **10** (1969)
240. M.H. Rider, in *SPWLA 19th Annual Logging Symposium* (1978)
241. A. Dumont, M. Kubacsi, J.L. Chardac, in *SPWLA 28th Annual Logging Symposium* (1987)

242. R.L. Kleinberg, W.C. Chew, E.Y. Chow, B. Clark, D.D. Griffin, SPE Form. Eval. **3**, 733 (1988)
243. J.P. Delhomme, T. Pilenko, E. Cheruvier, R. Cull, J. Petrol. Technol. **40**, 180 (1988)
244. B.E. Hornby, in *SEG Annual Meeting of the Society of Exploration Geophysicists* (1988)
245. V. Hepp, Log Anal. **33** (1992)
246. S.E. Prensky, Advances in borehole imaging technology and applications. Geol. Soc. Spec. Publ. **159**, 1 (1999)
247. R. Cheung, D. Pittman, A. Hayman, R. Laronga, R. Vessereau, A. Ounadjela, O. Desport, S. Hansen, M. Lamb, T. Borbas, B. Wendt, in *SPWLA 42nd Annual Logging Symposium* (2001)
248. L. Chabot, D.C. Henley, R.J. Brown, J.C. Bancroft, in *SEG Annual Meeting of the Society of Exploration Geophysicists* (2001)
249. Y. Zheng, X. Tang, in *SEG Annual Meeting of the Society of Exploration Geophysicists* (2005)
250. L.T. Bourke, D.J. Prosser, in *SPWLA 51st Annual Logging Symposium* (2010)
251. J. Prosser, L. Bourke, PESA News **113**, 26 (2011)
252. J. Prosser, L. Bourke, PESA News **114**, 78 (2011)
253. R. Bloemenkamp, T. Zhang, L. Comparon, R. Laronga, S. Yang, S. Marpaung, E.M. Guinois, G. Valley, P. Vessereau, E. Shalaby, B. Li, in *SPWLA 55th Annual Logging Symposium* (2014)
254. M.P. Ekstrom, C.A. Dahan, M.Y. Chen, P.M. Lloyd, D.J. Rossi, in *SPWLA 27th Annual Logging Symposium* (1986)
255. L.M. Grace, S.M. Luthi, R.G. Pirie, in *SPE Annual Technical Conference and Exhibition* (1986)
256. K.A. Safinya, P. Le Lan, M. Villegas, P.S. Cheung, in *SPE Annual Technical Conference and Exhibition* (1991)
257. D. Seller, G. King, D. Eubanks, in *SPWLA 35th Annual Logging Symposium* (1994)
258. J. Lofts, M. Evans, M. Pavlovic, S. Dymmock, Petrophysics **44** (2003)
259. P. Gaillot, T. Brewer, P. Pezard, E.C. Yeh, Sci. Drill. **5**, 1 (2007)
260. J. Brown, B. Davis, K. Gawankar, A. Kumar, B. Li, C.K. Miller, R. Laronga, P. Schlicht, Oilfield Rev. **27**, 4 (2015)
261. J. Zemanek, R.L. Caldwell, E.E. Glenn Jr., S.V. Holcomb, L.J. Norton, A.J.D. Straus, J. Petrol. Technol. **21**, 762 (1969)
262. D. Seller, C. Edmiston, O. Torres, J. Goetz, in *SPWLA 31st Annual Logging Symposium* (1990)
263. A.J. Hayman, P. Parent, P. Cheung, P. Verges, SPE Prod. Facil. **13**, 5 (1998)
264. Z.S. Leonard, in *IEEE Ultrasonics Symposium* (2016)
265. Z.S. Leonard, S. Rahman, R.R. Steinsiek, in *Offshore Technology Conference* (2017)
266. G.Aithoff, A. Arian, A.B. Kavaipatti, G.L. Varsamis, L.T. Wisniewski, in *SPWLA 39th Annual Logging Symposium* (1998)
267. C, Maeso, I. Tribe, in *SPE Annual Technical Conference and Exhibition* (2001)
268. M. Inaba, D. McCormick, T. Mikalsen, M. Nishi, J. Rasmus, H. Rohler, I. Tribe, Oilfield Rev. **15**, 24 (2003)
269. B. Elahifar, A. Esmaeili, R.K. Fruhwirth, G. Thonhauser, in *SPE Asia Pacific Oil and Gas Conference and Exhibition* (2012)
270. B. Elahifar, A. Esmaeili, R.K. Fruhwirth, G. Thonhauser, in *International Petroleum Technology Conference* (2013)
271. T.C. Sheives, L.N. Tello, V.E. Maki Jr., T.E. Standley, T.J. Blankinship, in *SPE Annual Technical Conference and Exhibition* (1986)
272. K.J. Goodwin, SPE Prod. Eng. **7**, 280 (1992)
273. G. Uswak, N. McLafferty, J. Can. Petrol. Technol. **34** (1995)
274. C.W. Morris, L. Sabbagh, R. Wydrinski, J.L. Hupp, R. van Kuijk, B. Froelich, in *SPE/IADC Drilling Conference* (2007)

275. Y. Guo, J. Qi, Y.S. Zheng, S. Taoutaou, R. Wang, Y. An, H. Guo, in *SPE/IATMI Asia Pacific Oil & Gas Conference and Exhibition* (2015)
276. S. Thierry, C. Klieber, M. Lemarenko, T. Brill, J.L. Le Calvez, F. Mege, T. Barrou, M. Lavrentiev, K. Constable, in *SPWLA 58th Annual Logging Symposium* (2017)
277. C. Klieber, A. Timonin, K. Singh, J.-L. Le Calvez, S. Thierry, M. Lemarenko, T. Brill, in *SPE Annual Caspian Technical Conference and Exhibition* (2016)
278. M. Lebourg, R.Q. Fields, C.A. Doh, Trans. AIME **210**, 260 (1957)
279. K.M. Conrad, in *SPE Drilling and Production Practices Conference* (1962)
280. A.L. Schultz, W.T. Bell, H.J. Urbanosky, J. Petrol. Technol. **27**, 1331 (1975)
281. J.J. Smolen, L.R. Litsey, J. Petrol. Technol. **31**, 25 (1979)
282. N. Colley, T. Ireland, P. Reignier, S. Richardson, J. Joseph, R. Zimmerman, A. Hastings, Oilfield Rev. **4**, 58 (1992)
283. R. Badry, E. Head, C. Morris, I. Traboulay, in *SPWLA 34th Annual Logging Symposium* (1993)
284. R. Badry, Oilfield Rev. **6**, 21 (1994)
285. J. Michaels, M. Moody, T. Shwe, in *SPE Annual Technical Conference and Exhibition* (1995)
286. C. Ayan, A. Douglas, F.J. Kuchuk, A revolution in reservoir characterization. MEARR **17**, 43 (1996)
287. A. Crombie, F. Halford, M. Hashem, R. McNeil, E.C. Thomas, G. Melbourne, O.C. Mullins, Oilfield Rev. **10**, 26 (1998)
288. A. Van Dusen, S. Williams, F.H. Fadnes, J. Irvine-Fortescue, in *SPE Annual Technical Conference and Exhibition* (2000)
289. R. Akkurt, M. Bowcock, J. Davies, C. Del Campo, B. Hill, S. Joshi, D. Kundu, S. Kumar, M. O'Keefe, M. Samir, J. Tarvin, Oilfield Rev. **18**, 4 (2006)
290. J. Creek, M.B. Cribbs, C. Dong, O.C. Mullins, H. Elshahawi, P. Hegeman, M. O'Keefe, K. Peters, J.Y. Zuo, Oilfield Rev. **21**, 38 (2009)
291. A. Cartellieri, J. Pragt, M. Meister, in *SPWLA 52nd Annual Logging Symposium* (2011)
292. A. Cartellieri, J. Pragt, F. Galvan Sanchez, M. Meister, in *SPE Annual Technical Conference and Exhibition* (2013)
293. T. Solbakk, H.E. Riise, E.A. Strøm, J. House, K.O. Eriksen, G. Digranes, in *SPE Bergen One Day Seminar* (2016)
294. D. Correa, A. Barbosa, in *SPE Latin America and Caribbean Petroleum Engineering Conference* (2017)
295. A. Cartellieri, T. Kischkat, S. Sroka, M. Meister, in *SPE/IADC Drilling Conference and Exhibition* (2017)
296. B.A. Hardage, Lead. Edge **4**, 59 (1985)
297. J.H. Justice, R. Hinds, A.F. Stirbys, in *Offshore Technology Conference* (1984)
298. L.J. Zimmerman, S.T. Chen, Geophysics **58**, 134 (1993)
299. B.E. Hornby, J. Yu, J.A. Sharp, A. Ray, Y. Quist, C. Regone, Lead. Edge **25**, 446 (2006)
300. P. Breton, S. Crepin, J.C. Perrin, C. Esmersoy, A. Hawthorn, R. Meehan, W. Underhill, B. Frignet, J. Haldorsen, T. Harrold, S. Raikes, Oilfield Rev. **14**, 32 (2002)
301. G. Hernandez, M. Casares, R. Perez, O. Barrios, R. Bautista, R.J. Brewer, J. Torne, in *48th Annual Logging Symposium* (2007)
302. S.J. Emsley, P. Shiner, N. Enescu, A. Beccacini, C. Cosma, Geol. Soc. Spec. Publ. **270**, 83 (2007)
303. J. Blackburn, J. Daniels, S. Dingwall, G. Hampden-Smith, S. Leaney, J. Le Calvez, L. Nutt, H. Menkiti, A. Sanchez, M. Schinelli, Oilfield Rev. **19**, 20 (2007)
304. P. Bulant, L. Klimeš, Geophysics **73**, U19 (2008)
305. C. Lange, S. Katsuko, in *Abu Dhabi International Petroleum Exhibition and Conference* (2015)
306. J.L. Arroyo, P. Breton, H. Dijkerman, S. Dingwall, R. Guerra, R. Hope, B. Hornby, M. Williams, R.R. Jimenez, T. Lastennet, J. Tulett, Oilfield Rev. **14**, 32 (2002)
307. G. Bernasconi, M. Vassallo, IEEE Trans. Geosci. Remote **41**, 687 (2003)

308. C. Naville, S. Serbutoviez, A. Throo, O. Vincké, F. Cecconi, Oil Gas Sci. Technol. **59**, 371 (2004)
309. A. Anchliya, in *SPE Europec/EAGE Annual Conference and Exhibition* (2006)
310. A. Ahmed, A. Soh, L. Morris, T. McMahon, M. Fernandez, in *IADC/SPE Asia Pacific Drilling Technology Conference* (2016)
311. J.W. Rector III, B.P. Marion, Geophysics **56**, 628 (1991)
312. R. Meehan, L. Nutt, N. Dutta, J. Menzies, in *SPE/IADC Drilling Conference* (1996)
313. D. Denney, J. Petrol. Technol. **52**, 21 (2000)
314. B. Hardage, Search and Discovery Article #40411 (2009). Available online at http://www.searchanddiscovery.com/documents/2009/40411hardage/index.htm
315. F. Poletto, F. Miranda, P. Corubolo, A. Schleifer, P. Comelli, Geophys. Prospect. **62**, 702 (2014)

Chapter 3
Instruments for Well Navigation and Drilling Optimization Evaluation

Abstract This chapter provides an introduction to drilling directional and horizontal wells and the two most common methods, sliding/rotating drilling with a bent-house hydraulic motor and rotary steerable systems (RSSs), utilized to accurately navigate a well to a target reservoir. At the heart of well navigation is a system known as measurement-while-drilling (MWD), which consists of magnetometers and accelerometers to measure the azimuth and inclination, respectively. A discussion on drilling dynamics sensors such as vibration, shock, downhole weight-on-bit (WOB), torque-on-bit (TOB) and drill bit revolutions per minute (RPM) to optimize drill bit penetration through rock formations is provided towards the end of the chapter.

3.1 Directional and Horizontal Wells

3.1.1 Principles of Downhole Navigation

Navigating a well is a complex procedure as the driller cannot see below the surface. As shown in Fig. 3.1, directional drilling is used to access reservoirs that are below populated areas or areas inaccessible from the surface such as reservoirs below mountains, perform side-tracks when a well cannot be drilled further vertically due to obstructions in the well or problems encountered due to unstable geological formations, access several types of resources from a single vertical well, such as oil, gas and water, and to explore for additional producing horizons in adjacent sectors of the field to maximize production [1–6]. Directional wells are also drilled to avoid collisions when there are several wells drilled on the same platform or in a field.

Drilling a directional well is a 3D process, as shown in Fig. 3.2, and modern directional technology allows a 3D well path of nearly any geometry to be drilled in order to reach drilling objectives [7–12]. A directional well is drilled by deviating from a vertical well path and then accurately guiding the well to a predefined target, typically a hydrocarbon reservoir [13, 14]. The key parameter required for successful navigation is the position of the drill bit at any given time, which is

C. P. Gooneratne et al., *Instruments, Measurement Principles and Communication Technologies for Downhole Drilling Environments*, Smart Sensors, Measurement and Instrumentation 32, https://doi.org/10.1007/978-3-030-04900-3_3

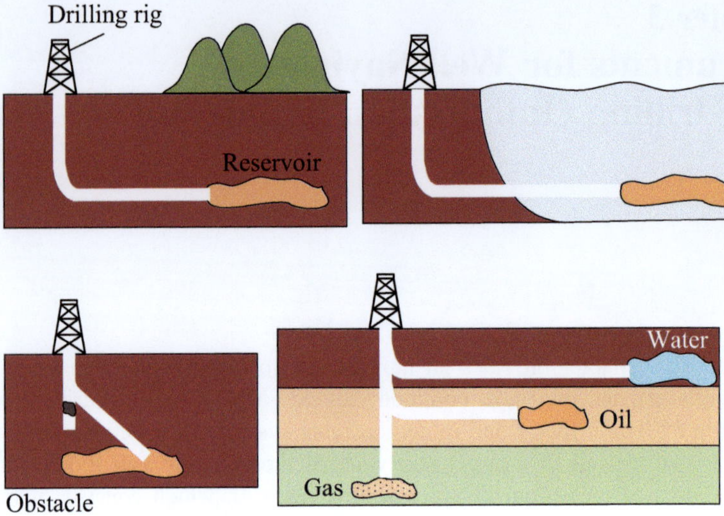

Fig. 3.1 Directional drilling to access downhlole reservoirs

calculated in terms of azimuth and inclination as shown in Fig. 3.2. The azimuth is the angle measured in a clockwise direction from the magnetic north to the projection of the well trajectory on the horizontal plane, where magnetic north and true north is separated by a declination angle that is dependent on the geographical location the well is drilled [15]. The inclination is the angle the well deviates from the vertical direction, with the vertical direction defined by the Earth's gravitational field. A 0° inclination signifies a vertical well.

The earliest wellbore position surveying methods, such as lowering an acid bottle into a well to etch an acid ring on the bottle and the Totco mechanical drifter, only measured the inclination [16]. These methods were followed by the magnetic single and multi-shot survey tools, which were the first tools to measure both azimuth and inclination. These tools measure the magnetic north with a compass and the inclination with an inclinator, on a disk, and the camera takes a photograph of the disk based on the electronic timer [17]. However, the wellbore survey data is not real-time since the single and multi-shot tools are run in the well by wireline (WL) or dropped down the drillstring assembly, so they have to be retrieved to obtain the survey data. Once the survey data is obtained several methods are available to guide the well to a deviated path. These include setting whipstocks, jetting tools, and a geometry of components in a passive bottomhole assembly designed to build, drop or maintain inclination even though these methods have little influence in controlling azimuth [5, 13, 14, 16, 18]. However, the development of a downhole mud motor and the advent of a downhole telemetry method for data communication called mud pulse telemetry, which is discussed in Chap. 5, shifted the paradigm of drilling directional wells onto a real-time navigation platform called measurement-while-drilling (MWD), where wells could be steered in any direction, while drilling, in real-time.

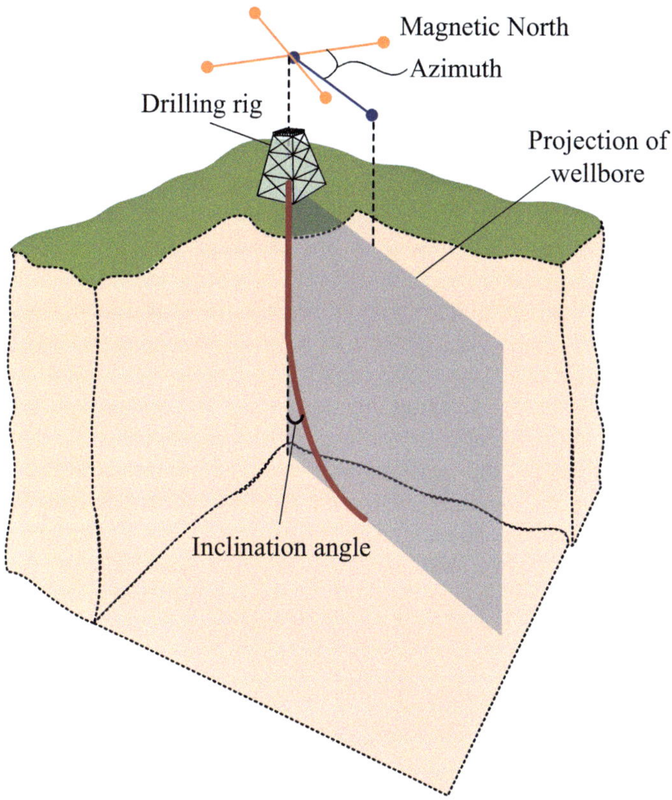

Fig. 3.2 Azimuth and inclination when drilling a directional well [6]

3.2 Measurement-While-Drilling and Navigation Instruments and Methods

3.2.1 Sensors for Downhole Navigation

The key drive behind the technological evolution of MWD was the need to navigate and steer a well in real-time, onto any desired path or trajectory, so that virtually any hydrocarbon reservoir downhole could be accessed from the surface, and to increase the efficiency of multi-well pads/locations especially in offshore environments. Measurement-while-drilling technology also enabled logging instruments, which were previously run by WL, to transmit real-time formation evaluation data to the surface for analysis [19, 20]. Commercial MWD tools arrived in the 1970s, and while early MWD units only had a downhole motor, directional instruments and the telemetry unit, today this basic unit is typically connected to other instruments such as vibration, pressure and temperature instruments, and formation evaluation instruments such as gamma ray, resistivity and neutron-porosity instruments [9, 11, 12, 21–24]. In this review we define MWD as, any real-time downhole measurement except formation evaluation measurements and telemetry methods, since the former was presented in Chap. 2 and the latter will be presented in Chap. 5.

MWD tools use triaxial fluxgate magnetometers (FGMs) and triaxial accelerometers, installed around 50 ft above the drill bit, as directional sensors [6]. These sensors have excellent temperature stability, low noise characteristics, low power consumption and can be constructed easily according to well established design principles at low cost. Most magnetometers and accelerometers used in MWD tools can measure azimuth and toolface with an accuracy of 1° and inclination with 0.1° accuracy. Figure 3.3 shows a schematic of triaxial magnetometers and accelerometers mounted inside a drill pipe. The earth's magnetic field has a different strength and orientation at every location on earth and the magnetometers measure the strength of the magnetic field along three orthogonal axes. Similarly the accelerometers measure the gravitational field along three orthogonal axes to determine the inclination of the well. Figure 3.4 shows the relationship between the drill bit, toolface, azimuth and inclination, where the toolface is the angle the drill bit rotates on the drilling plane from an initial reference point. The gravitational and magnetic fields must be present in at least 2 sensors of each kind for deriving a toolface measurement. If the well is vertical (0° inclination), magnetometers measure the toolface angle since the gravitational field is constant and present only across the z-axis accelerometer. If the well is parallel to the Earth's magnetic field, the toolface can only be measured by accelerometers. Apart from these two situations, the

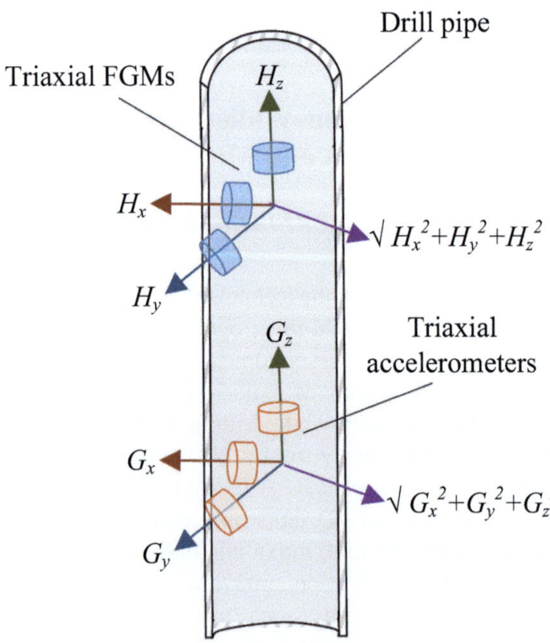

Fig. 3.3 Triaxial fluxgate magnetometers and triaxial accelerometers mounted inside a drill pipe to measure the azimuth and inclination of a well while drilling [6]

toolface is measured by both the magnetometers and the accelerometers. In most instances a crossover from magnetic toolface to gravitational toolface occurs at angles from 3° to 5°. Magnetic field measurements by the magnetometers, inclination measurements by the accelerometers, and toolface measurements can be used to calculate the position P of the bit in a directional well at any time using the following equation [25].

$$P = \arctan\left(\frac{-(H_x \sin\phi + H_y \cos\phi)}{H_z \sin\theta + \cos\theta(H_x \cos\phi - H_y \sin\phi)}\right) \tag{3.1}$$

and

$$\theta = \arctan\left(\sqrt{\frac{G_x^2 + G_y^2}{G_z^2}}\right) \tag{3.2}$$

and

$$\phi = \arctan\left(-\frac{G_y}{G_x}\right) \tag{3.3}$$

where H_x, H_y and H_z are the orthogonal magnetic field vectors, G_x, G_y, G_z are the orthogonal gravitational vectors, θ is the inclination and ϕ is the toolface angle.

Azimuth, inclination and toolface are obtained at predetermined survey points along the well and combined based on mathematical assumptions to obtain the well trajectory, which is then transferred to the surface so the driller knows the exact location of the well.

3.2.2 *Downhole Navigation Methods*

3.2.2.1 Sliding and Rotating

Today, the most common method of drilling a directional well is by the sliding and rotating technique [26–28]. Figure 3.5 shows a drillstring assembly configuration used to implement this technique. As shown in Fig. 3.5a, the MWD unit consists of: the directional sensors; the bent-housing motor comprising a hydraulic motor driven by the drilling fluid flow, where the bend of the bent-house motor is adjusted according to the angle of the directional well being drilled and is typically between 0° and 2°, but can be high as 3°; and the stabilizers, which are used to control contact with the wellbore and also to act as a lever so that side force is generated at the bit. In sliding mode, the hydraulic bent-housing motor is oriented in the desired direction and powered by the drilling fluid flow that flows through the drillstring assembly, driving the drill bit to build or drop angle or turn while the drillstring

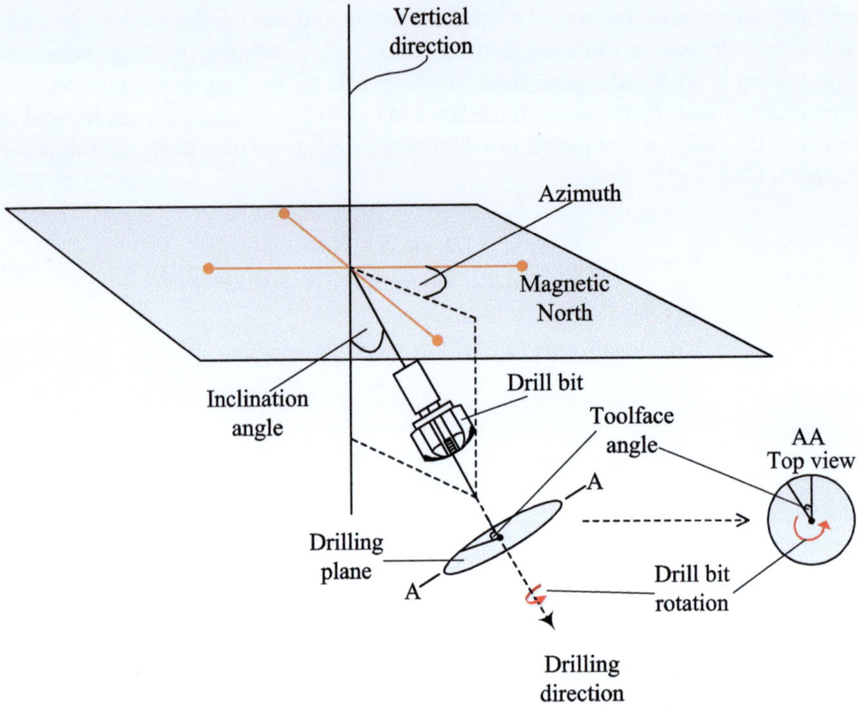

Fig. 3.4 The azimuth of a directional well is the deviation from the magnetic north and the inclination is the deviation from the vertical direction of the well. The toolface is the angle the drill bit rotates on the drilling plane from an initial reference point [6]

assembly above the drill bit remains non-rotating, as shown in Fig. 3.5b. Once the desired target trajectory is attained, the entire drillstring assembly and the drill bit rotate to drill ahead in rotating mode, as shown in Fig. 3.5c.

3.2.2.2 Rotary Steerable System

Rotary steerable system (RSS) technology was first introduced two decades ago as an alternative to the sliding and rotating method of directional drilling, since the sliding motion to establish a trajectory is significantly slower than the rotating mode resulting in a time-consuming operation [28–31]. The RSS technology allows the constant rotation of the entire drillstring assembly while drilling a directional well, thereby increasing the drilling speed and reducing the time to drill a well. While RSS technology saves money by drilling faster it is also more expensive to run in a well than a sliding/rotating system. Therefore, cost-effective planning and optimization must be performed taking into account factors such as production potential of the well, type of rock formation, depth and angle of a well, drill bit compatibility and expected wellbore problems.

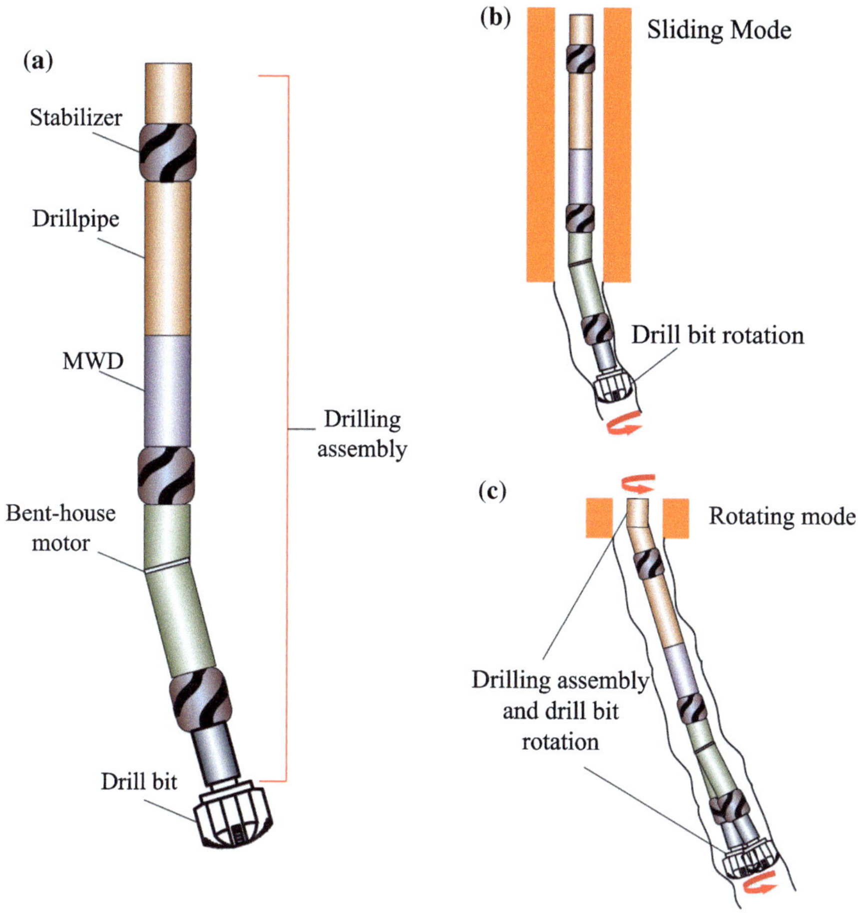

Fig. 3.5 **a** Drilling assembly for directional drilling with an MWD unit consisting of FGMs and accelerometers to obtain azimuth and inclination measurements of the well, a bent-housing motor, which initiates the trajectory of the well being drilled, as shown in **b**, and stabilizers that allow side force to be generated at the bit. Once the desired trajectory is obtained the whole drilling assembly and the bent-motor drills ahead as shown in **c** [6]

The two main RSS technologies are point-the-bit, as shown in Fig. 3.6a, and push-the-bit, as shown in Fig. 3.6b [32–38]. The point-the-bit RSS has a shaft that is bent relative to the rest of the drilling assembly, where the bend angle is controlled by the driller at the surface through commands sent downhole through telemetry, and the commands depend on the azimuth, inclination and toolface measurements sent up to the surface by MWD. The bend orientation is controlled by a servomotor that rotates at the same speed as the drillstring assembly but in the opposite direction so that the toolface orientation remains non-rotating while the drillstring assembly rotates. To make the trajectory of the well change in the direction of the bend the reference stabilizer acts as a reference to create the

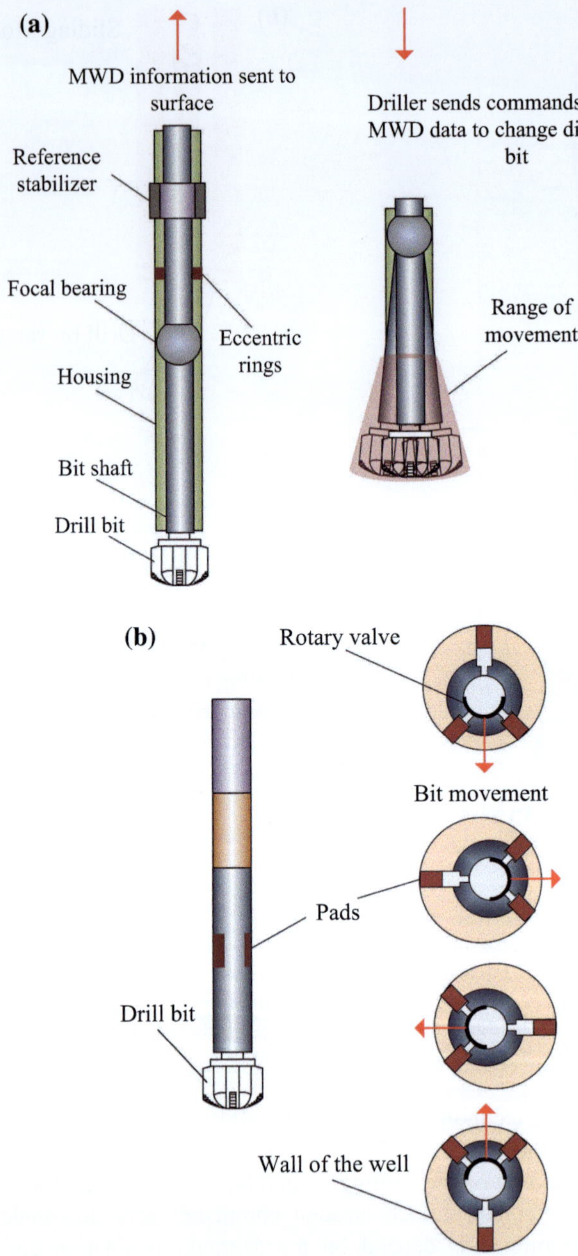

Fig. 3.6 **a** Point-the-bit system, where the bit shaft is bent relative to the rest of the drilling assembly. The bend orientation is controlled by a servomotor that rotates at the same speed as the drilling assembly but in the opposite direction so that the toolface orientation is non-rotating; **b** Push-the-bit system, where the pads are actuated by a flow and pushed out against the wall of the well being drilled to direct the drillstring assembly in a desired direction [6]

deflection within the shaft. The push-the-bit RSS has pads that are actuated by the drilling fluid flow to push out against the wall of the well being drilled and direct the drillstring in the desired trajectory. The drilling fluid flow is supplied to the pads by opening and closing a rotary valve and the timing and magnitude of pad actuation is controlled by a driller at the surface who sends commands downhole based on the directional data sent to the surface utilizing MWD telemetry.

3.2.2.3 Navigation Errors

The use of magnetometers influences the overall accuracy of the surveying process. Factors such as magnetic interference by drill string components and formation ore deposits influence magnetometer measurements. Since the measured azimuth reference is the magnetic north, any magnetic interference due to any field other than the Earth's magnetic field will influence the magnetometer measurements and the magnetic azimuth accuracy. Therefore, magnetometers are enclosed in non-magnetic drill pipes when run inside a well. The amount of drill pipe required to effectively isolate the magnetometers depends on the accuracy of the 'proposed azimuth' and inclination. As the accuracy depends on the signal levels present on the different sensors, vertical is usually not a difficult case and 90° or horizontal due east or west are the most challenging at locations where the Earth's magnetic field is nearly horizontal.

An ellipsoid of uncertainty (EOU) error model applicable to a basic MWD is used to determine the errors due to combined sensor accuracy [14, 16, 39, 40]. The well position uncertainty is visualized by using tools from different companies in directional drilling software developed to integrate with subsurface applications. Errors calculated at well survey stations placed a maximum of 100 ft apart are based on three main vectors, the azimuth, inclination and the depth of the well, while the toolface angle is also considered when modeling the propagation of errors. Errors arise from different sources and are statistically independent, cumulative and propagate in proportion to the distance from the origin, so the final location of the well can fall anywhere within the final position uncertainty, as shown in Fig. 3.7a. An error term describes the error source, which is any phenomenon that influences an MWD measurement. Therefore, an error model can be defined as a set of error terms that account for all the different error sources in an MWD tool such as sensor errors, errors due to steel in the drilling assembly near the MWD, directional sensor misalignments and magnetic dip and field strength uncertainties. Figure 3.7b shows the lateral dimension of an EOU is proportional to the azimuth error and high side dimension is proportional to the inclination error. When the azimuth is more accurate than the inclination the ellipse is thinner and when the inclination is more accurate than the azimuth the ellipse is more spread-out, which is more typical in the field. The addition of well depth as the third component of the error into the EOU results in an almond-like shape, elliptical in all three orthogonal planes. The most important application of the EOU model is the avoidance of well collision. It is also used to estimate and increase the probability of

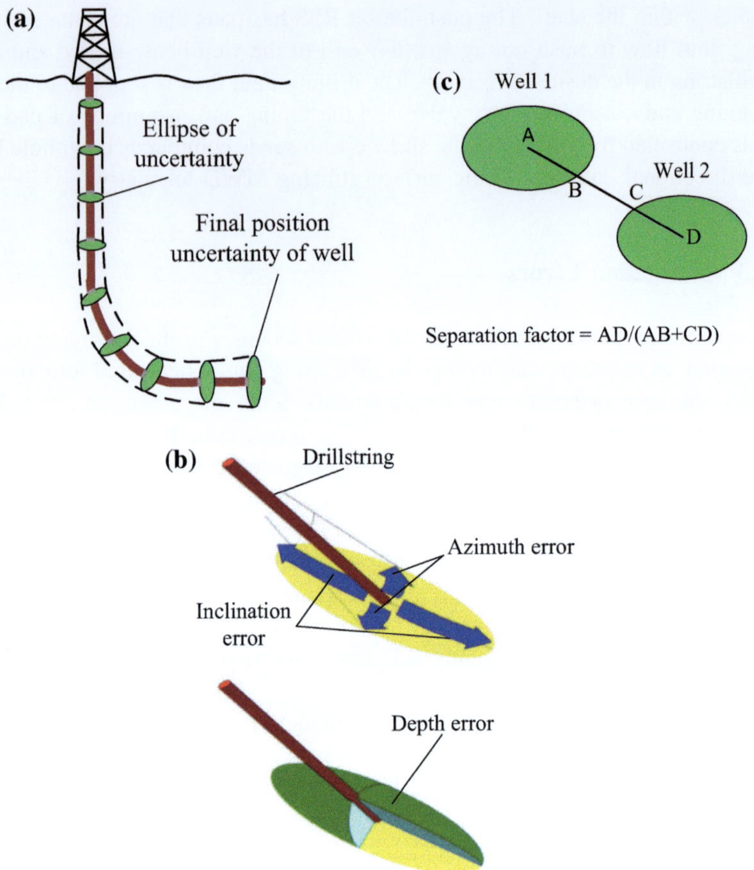

Fig. 3.7 **a** The ellipses of uncertainty gets larger as the depth from the surface increases since errors from different sources are statistically independent, are cumulative and errors propagate in proportion to how far you are from the origin; **b** The lateral dimension of the ellipse of uncertainty is proportional to the azimuth error and the high side dimension is proportional to the inclination error; **c** Calculation of separation factor between two wells to avoid well collision [6]

navigating inside a target reservoir, especially when the reservoir is thin. The simplest method of evaluating collision risk is by using the separation factor method shown in Fig. 3.7c, where the separation factor is calculated in a plane at right angles to the well that has already been drilled. Wells just touching is specified by a separation factor of 1 but this does not mean the wells will definitely collide since a probability with 2 or 3 standard deviations are used when determining the size of ellipses, even though it is worth noting that the probability of well collision is proportional to the overlapping region. There are also other methods of calculating the separation factor, which depend on the geometry of the subject well and the offset wells in the region.

Gyroscopes measure the rate of the earth's rotation, which does not change with time and is not affected by the presence of ferrous materials. Gyroscopes were originally run as single or multi-shot instruments but are now increasingly being used in MWD units while navigating wells. Gyro-while-drilling (GWD) can be performed in environments where there is significant magnetic interference because unlike magnetometers they are not affected by magnetic fields and ferromagnetic materials and do not require non-magnetic drill collars to mitigate magnetic interference [41–44]. However, gyroscopes are not as rugged as fluxgate magnetometers, have relatively high power consumption and are much more expensive. Moreover, there is continuous development to address compatibility issues with the currently available MWD EOU model, and ascertain if the rotating speed of the drillstring assembly introduces geographic reference errors and unaccountable measurement drifts that might affect the quality of the GWD survey.

3.2.3 Drilling Dynamics Sensors

The Drilling dynamics sensors such vibration, shock, downhole weight on bit (WOB), torque on bit (TOB) and drill bit revolutions per minute (RPM) are mounted on a special unit close to the bit, as shown in Fig. 3.8, to diagnose any issues related to the drilling mechanism and provide the driller at the surface information to optimize drill bit penetration through rock formations. Moreover, the real-time data provided by the drilling dynamics sensors, such as vibration and shock, can be correlated to downhole tool and instrument, drillstring and drill bit fatigue and failures and this information can be used to reduce these effects by modifying drilling parameters [45–49]. Triaxial accelerometers are utilized for measuring axial (z), lateral and torsional (x and y) vibrations, bending strain gauges may be placed parallel to the drillstring assembly for detecting axial forces and WOB, at 45° to the drillstring assembly for torsional forces for TOB, and magnetometers can be used to monitor the drill bit RPM. For example, high acceleration in the axial direction is linked with drill bit bouncing due to axial vibrations, high average acceleration in the lateral direction is related to bit whirl, which is caused by drillstring eccentricities resulting in centripetal forces during rotation, and separation between average and peak x and y accelerations is associated with stick-slip of the drill bit, which is due to torsional vibration of the drillstring resulting in alternating stops and intervals of high velocity rotation of the drill bit [50, 51]. The accelerometers have a typical range of 0–200 g with a resolution of 0.1 g and a frequency response up to 5000 Hz, the strain-gauges have a 2-axis bending range of 0–25,000 ft lb with a resolution of 150 ft lbs and a WOB tolerance of 90,000 lbs with a resolution 500 lbs. However, bending strain gauges are temperature sensitive and are therefore difficult to calibrate. Drilling dynamics sensors are mostly used in complex drilling environments, such as high impact and abrasive formations with interbedded intervals, and are also recognized as a key component in the drive towards drilling optimization and automation [52].

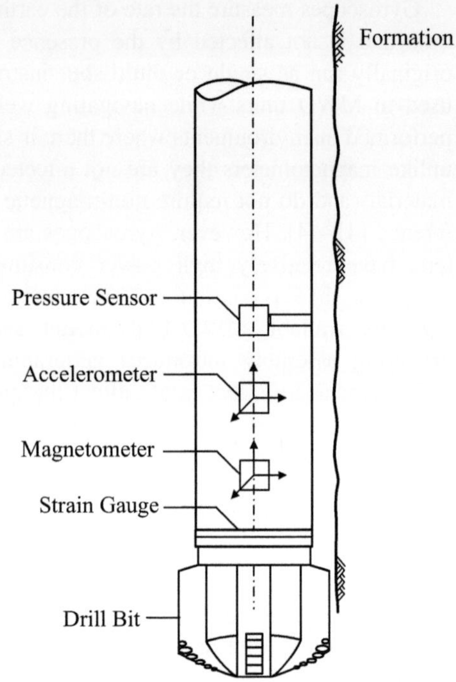

Fig. 3.8 Drilling dynamics unit, to measure vibration, shock, weight on bit, torque on bit, bit revolution speed, pressure and temperature, mounted close to the drill bit

Pressure and temperature measurements in wells provide information on the downhole pressure and temperature gradients, which can be utilized to select drilling and logging instruments based on their temperature/pressure tolerances, calibrate logging measurements, and develop and fine-tune recipes for designing drilling fluids, cement slurries and to employ other chemicals for balancing formation pore pressures, improve wellbore wall strength and addressing other wellbore construction problems. Formation pressure and temperature data can be obtained by quartz or strain-gauge pressure sensors and thermistors, respectively, as explained before during WL fluid sampling and testing. This data can then be used to estimate the temperature and pressure gradients for the whole well with considerations for different types of formations.

Most of the downhole problems encountered while drilling is associated with operating outside the low/high boundaries of pressure, which is defined by the formation pore, collapse and fracture pressures. The formation pressure and temperature gradients for a well is based on measurements obtained by WL logging, from inside the formation, in a static environment. The bottomhole pressures and temperatures while drilling can deviate significantly from the static values calculated from the pressure and temperature gradients, due to the friction caused by the circulating drilling fluid with rock formation cuttings, and the possible variation of drilling fluid density with pressure (compressibility) and temperature (thermal expansion). Measurement of the bottomhole pressure and temperature becomes very important when the margin between formation pore pressure and the fracture gradient, known as the drilling window, is very small. There are many bottomhole hydraulic analysis software packages that aim to predict the bottomhole pressure and temperature but these values are based on a number of assumptions that in most cases do not accurately reproduce the actual real-time drilling environment [53, 54]. Pressure-while-drilling (PWD) instruments, as shown in Fig. 3.8, were developed in the late 1990s to directly measure the bottomhole pressure inside and outside the drillstring assembly, and remove uncertainties created by the hydraulic models [55–60]. PWDs have quartz, sapphire or strain-gauge type sensors mounted inside and outside the drillstring to measure the pressure inside the drillstring assembly and outside in the annulus [61, 62]. The sensors can measure pressure up to 20,000 psi with a resolution of 1 psi. PWD is used in almost all wells as part of a standard MWD package, especially in deepwater wells where inherently unstable formations are prone to unpredictable pressures changes, and large natural fractures lead to lost circulation of drilling fluids. One method to drill in these challenging conditions, which present very narrow drilling windows, is by managed pressure drilling, which uses surface equipment to manage downhole pressures. An influx of formation fluids into the well will result in an increase in bottomhole pressure in the annulus, and in such a situation, a PWD instrument provides real-time pressure recordings to the surface so the driller can immediately use a choke to balance the sudden increase in the annulus pressure. Therefore, the PWD aids fast decision making to produce a rapid well control response, where failure to do so may result in the well flowing at the surface [63].

Temperature sensors are generally built into the PWD instrument to monitor the temperature inside the drillstring assembly and outside in the annulus. Temperature sensors are able to record a maximum temperature of 230 °C with a resolution of 0.5 °C. While accurate prediction of bottomhole temperature contributes to an efficient drilling operation as a pressure sensor does, it also plays an important role when designing cement recipes, by optimizing cement thickening times, compressive strength, rheological properties and set time recipes, to permanently seal casing to a wellbore. Furthermore, bottomhole temperature is a key parameter for designing resin-based lost circulation recipes, which start off as viscous liquids and harden with temperature, to plug fractures in a rock formation.

References

1. L.E. Pendleton, *Archie Conference on Reservoir Definition and Description* (1991)
2. I.B. Ishak, R.P. Steele, R.C. Macaulay, P.M. Stephenson, S.M. Al Mantheri, *SPE Middle East Oil Show* (1995)
3. D.A. Gust, *Annual Technical Meeting of the Petroleum Society of Canada* (1995)
4. W.C. Maurer, J. Can. Petrl. Technol. **34** (1995)
5. K. Mantle, Oilfield Rev. **25**, 54 (2013)
6. C.P. Gooneratne, B. Li, T.E. Moellendick, Sensors **17**, 2384 (2017)
7. R. Mottahedeh, *Canadian International Petroleum Conference* (2005)
8. R. Mottahedeh, J. Can. Petrl. Technol. **47** (2008)

9. S.H. Al-Suwaidi, S. Lyngra, I. Roberts, J. Al-Hussain, I. Pasaribu, A.S. Laota, S. Hutabarat, *SPE Saudi Arabia Section Technical Symposium and Exhibition* (2013)
10. A.M.A. Al-Maskeen, R.R. Sung, S.S. Ali, *SPE Middle East Oil and Gas Show and Conference* (2013)
11. S. Jamaluddin, L. Hanalim, W. Suwarlan, G.I. Santoso, D. Permanasari, K.A. Alang, *International Petroleum Technology Conference* (2014)
12. D. Cedeno, A. Alvarez, J. Fuentes, C. Portilla, V. Machare, R. Erazo, K. Escobar, R.E. Cedeño, *SPE Latin America and Caribbean Petroleum Engineering Conference* (2017)
13. A.T. Bourgoyne Jr., K.K. Millheim, M.E. Chenevert, F.S. Young Jr, *Applied Drilling Engineering* (SPE, 1991)
14. B. Mitchell, *Advanced Oilwell Drilling Engineering—Handbook and Computer Programs* (SPE, 1995)
15. J.W. Wright, *SPE/IADC Drilling Conference* (1988)
16. A. Buchanan, C.A. Finn, J.J. Love, E.W. Worthington, F. Lawson, S. Maus, S. Okewunmi, B. Poedjono, Oilfield Rev. **25**, 32 (2013)
17. J.L. Thorgood, D.R. Knott, SPE Drill. Eng. **5**, 209 (1990)
18. T. Ma, P. Chen, J. Zhao, Geomech. Geophys. Geo-Energy Geo-Resour. **2**, 365 (2016)
19. D.R. Tanguy, W.A. Zoeller, *SPE Annual Technical Conference and Exhibition* (1981)
20. C.A. Martin, R.M. Philo, D.P. Decker, T.M. Burgess, *SPE/IADC Drilling Conference* (1994)
21. W. Gravley, J. Petrol. Technol. **35**, 1439 (1983)
22. J.E. Fontenot, J. Petrol. Technol. **38**, 128 (1986)
23. A.A. Al-Fawwaz, N.M. Al-Musharfi, P.J. Butt, A. Fareed, *SPE/IADC Middle East Drilling and Technology Conference* (2007)
24. F. Makohl, R. Jurgens, *SPE/IADC Drilling Conference* (1986)
25. M.K. Russel, A.W. Russel, *Surveying of Boreholes*. U.S. Patent 4,163,324 (1979)
26. E. Maidla, M. Haci, *SPE/IADC/SPE Drilling Conference* (2004)
27. E. Maidla, M. Haci, S. Jones, M. Cluchey, M. Alexander, T. Warren, *SPE/IADC Drilling Conference* (2005)
28. E. Felczak, A. Torre, N.D. Godwin, K. Mantle, S. Naganathan, R. Hawkins, K. Li, S. Jones, F. Slayden, Oilfield Rev. **23**, 36 (2011)
29. K. McMillin, Offshore Magazine (1999). Available online at https://www.offshore-mag.com/articles/print/volume-59/issue-2/departments/drilling-production/rotary-steerable-systems-creating-niche-in-extended-reach-drilling.html
30. M. Mims, World Oil **220**, 40 (1999)
31. G. Downton, T.S. Klausen, A. Hendricks, D. Pafities, Oilfield Rev. **26**, 18 (2000)
32. S. Schaaf, C.R. Mallary, D. Pafitis, SPE Annual Technical Conference and Exhibition (2000)
33. S. Schaaf, D. Pafitis, *SPE/IADC Drilling Conference* (2001)
34. D. Stroud, M. Russell, S. Peach, *SPE Annual Technical Conference and Exhibition* (2003)
35. B. Cox, L.A. Romo, B.D. Champion, K.D. Card, S.P. Barton, Z. Maung, *SPE/IADC Drilling Conference* (2007)
36. H. Al Yami, A. Kubaisi, K. Nawaz, A.H. Awan, J.K. Verma, S. Ganda, *SPE Asia Pacific Oil and Gas Conference and Exhibition* (2008)
37. J. Sugiura, *SPE Asia Pacific Oil and Gas Conference and Exhibition* (2008)
38. V. Kretsul, S. Dymov, P. Dobrokhleb, A. Milenkiy, O. Tarasov, *SPE Russian Petroleum Technology Conference* (2015)
39. H.S. Williamson, SPE Drill. Complet. **15**, 221 (2000)
40. A. Jamieson, Introduction to wellbore surveying. Available online at http://www.uhi.ac.uk/en/researchenterprise/energy/wellbore-positioning-download
41. T. Torkildsen, S.T. Havardstein, J.L. Weston, and R. Ekseth, SPE Drill. Complet. **23**, 5 (2008)
42. J.L. Weston, A.G. Ledroz, R. Ekseth, SPE Drill. Complet. **29**, 295 (2014)
43. A.E. Okwu, E.B. Hughes, *SPE Nigeria Annual International Conference and Exhibition* (2015)
44. R. Lowdon, J. Igarashi, D. Sargin, *SPE Annual Technical Conference and Exhibition* (2017)

45. J.R. Bailey, C.C. Elsborg, R.W. James, P. Pastusek, M.T. Prim, W.W. Watson, SPE Drill. Complet. **28**, 350 (2013)
46. B. Saldivar, I. Boussaada, H. Mounier, S. Mondie, S.I. Niculescu, IFAC P. Ser. **47**, 5169 (2014)
47. A. Ghasemloonia, D.G. Rideout, S.D. Butt, J. Petrol. Sci. Eng. **131**, 150 (2015)
48. G. Dong, P. Chen, Shock and Vibration (2016). Available online at http://dx.doi.org/10.1155/2016/7418635
49. O. Akimov, A. Hohl, H. Oueslati, H. Reckmann, *SPE/IADC Middle East Drilling Technology Conference and Exhibition* (2018)
50. P. Centala, V. Challa, B. Durairajan, R. Meehan, L. Paez, U. Partin, S. Segal, S. Wu, I. Garrett, B. Teggart, N. Tetley, Oilfield Rev. **23**, 4 (2011)
51. R.I. Leine, D.H. Van Campen, W.J.G. Keultjes, J. Vib. Acoust. **124**, 209 (2002)
52. E. Cayeux, B. Daireaux, E.W. Dvergsnes, and F. Florence, SPE Drill. Complet. **29**, 236 (2014)
53. I.M. Kutasov, L.V. Eppelbaum, *Proceedings of the World Geothermal Congress* (2015), pp. 19–25
54. D. Denney, J. Petrol. Technol. **53**, 33 (2001)
55. C.D. Ward, E. Andreassen, *SPE/IADC Drilling Conference* (1997)
56. M. Hutchinson, I. Rezmer-Cooper, *SPE Annual Technical Conference and Exhibition* (1998)
57. W. Aldred, J. Cook, P. Bern, B. Carpenter, M. Hutchinson, J. Lovell, I. Rezmer-Cooper, P.C. Leder, Oilfield Rev. **10**, 40 (1998)
58. I. Rezmer-Cooper, T. Bratton, H. Krabbe, SPE Drill. Complet. **16**, 35 (2001)
59. S.G. LaPierre, G. Courville, J. Song, *SPE/IADC Drilling Conference* (2006)
60. E. Mammadov, N. Osayande, S. Sephton, J. Risi, *SPE Digital Energy Conference and Exhibition* (2015)
61. E.P. EerNisse, R.B. Wiggins, IEEE Sens. J. **1**, 79 (2001)
62. Y. Barriol, K.S. Glaser, J. Pop, B. Bartman, R. Corbiell, K.O. Eriksen, H. Laastad, J. Laidlaw, Y. Manin, K. Morrison, C.M. Sayers, Oilfield Rev. **17**, 22 (2005)
63. A. Ennaifer, P. Giordano, S. Vannuffelen, B.A. Nilssen, I. Nwagbogu, A. Sooklal, C. Walden, Oilfield Rev. **26**, 32 (2014)

Chapter 4
Logging Tools in Hydrocarbon Wells

Abstract In this chapter we present the manner in which logging instruments are employed in hydrocarbon wells to optimize drilling and reach target reservoirs to recover the maximum volume of hydrocarbons. The methods seven different types of logging tools, resistivity, induction, dielectric, spectral azimuthal gamma, pulsed neutron, nuclear magnetic resonance, interpret downhole measurements and provide information about reservoir properties to the driller and the petrophysicist in real-time are explained.

4.1 Logging Instruments for Formation Evaluation

4.1.1 Resistivity Measurement Tool

A logging-while-drilling (LWD) propagation resistivity tool was utilized in a thin reservoir to aid in detecting approaching reservoir boundaries [1–4]. This ensures that the well is steered to be within the reservoir for maximum access to hydrocarbons. The resistivity of the reservoir is higher than the resistivity of the surrounding formations. Figure 4.1 shows the azimuthal resistivity tool utilized in the well with six transmitters, 3 upper (T1-3) and 3 lower (T4-6) transmitters, and 3 receivers (R1-3) tilted 45°. The longest transmitter-receiver spacing is 112 inches and the shortest is 16 inches [1–4]. This configuration of transmitters and receivers allows measurements at multiple depths of investigation (DOI) and the tilted receivers enable an azimuthally sensitive resistivity measurement that can be utilized to create images. As the drillstring rotates during drilling, the resistivity measurements scan the formation surrounding the wellbore into 32 azimuthal resistivity readings corresponding to the formation at varying DOI. Therefore, these measurements provide a 3D characterization of the near wellbore environment allowing precise recognition of laterally approaching formations from several feet away. The DOI refer to deep and shallow resistivity logs, where the shallow logs refer to measurements obtained at the wall of the wellbore, closest to the resistivity

C. P. Gooneratne et al., *Instruments, Measurement Principles and Communication Technologies for Downhole Drilling Environments*, Smart Sensors, Measurement and Instrumentation 32, https://doi.org/10.1007/978-3-030-04900-3_4

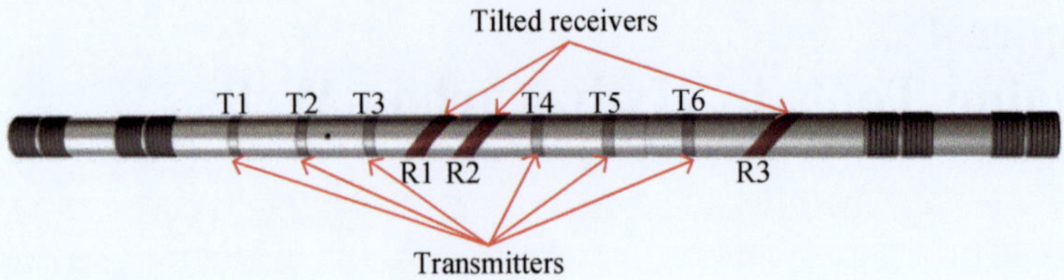

Fig. 4.1 The azimuthal propagation resistivity tool with six transmitters and three tilted receivers that provide azimuthal sensitivity. Shallow and deep measurements are enabled by the long and short transmitter/receiver spacings [1]

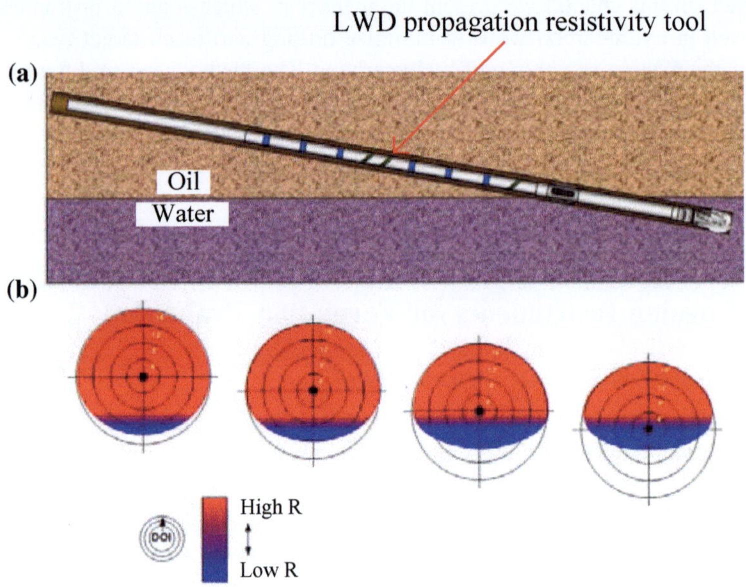

Fig. 4.2 The azimuthal propagation resistivity tool for geosteering in reservoir zones. **a** Tool moving from an oil to a water zone. **b** Azimuthal and radial resistivity plots reflecting the crossing of the oil-water boundary by an increase in the low resistivity portion of the plot (in blue) [1]

logging tool, and the deep logs refer to measurements obtained deep within the formation, furthest away from the logging tool and uncontaminated by the resistivity logging tool.

Figure 4.2 shows an example of azimuthal deep resistivity measurements changing as the tool approaches an oil/water contact [1]. Figure 4.2a shows the tool moving from an oil formation to a water formation, and Fig. 4.2b shows the azimuthal and radial resistivity measurement plots reflecting this by an increase in the low resistivity portion of the plot (in blue). However, formations are rarely flat and have irregular dips so two methods, as shown in Fig. 4.3, are used when steering the well to be within a reservoir [1]. One method known as reactive geosteering, as shown in Fig. 4.3a, relies

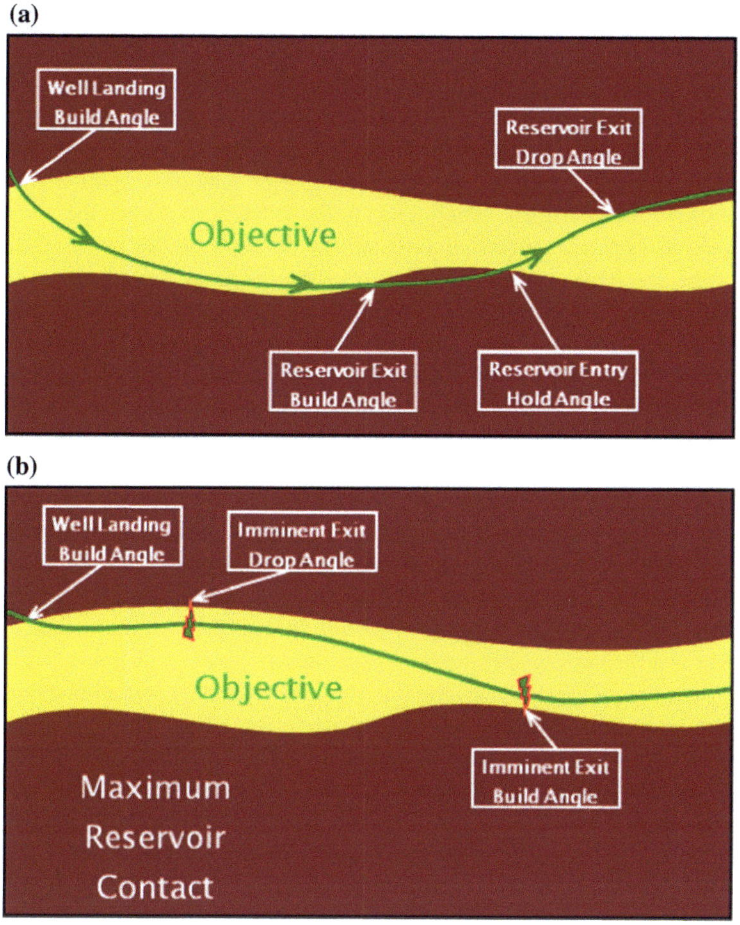

Fig. 4.3 **a** Reactive geosteering. **b** Proactive geosteering [1]

on logging sensor measurements to change the direction of the well path. However, in this method the well path is maneuvered back to the reservoir zone after intercepting the surrounding formation boundaries. Proactive geosteering, as shown in Fig. 4.3b, changes the direction of the well before intercepting the surrounding boundaries to be within the target area. Proactive geosteering makes use of deeper DOI tools to detect boundaries and adjust the well direction while reactive geosteering utilize shallower DOI tools. Generally both proactive and reactive geosteering are employed for optimal placement of wells and this depends on type, DOI and the angle of incidence between the drilling assembly and the surrounding geology. When the angle of incidence is low, deeper DOI sensors located further away from the drill bit will indicate changes in geology before the shallow DOI sensors located near the drill bit. On the other hand, the shallower sensors will detect changes in geology first when the angle of incidence is high.

Figure 4.4 shows the real-time response of the LWD propagation resistivity tool in a downhole geological setting defined within the total vertical depth (TVD) and the measured depth (MD) of the well path, where a high resistive reservoir (in yellow) is surrounded by a lower resistive formation (in green) as shown in track 1 [1]. Track 2 shows the azimuthal resistivity obtained by the tool. The up and down resistivities in track 3 refer to the upper and lower bins of azimuthal resistivity. Track 4 shows the geosignal, which refers to the difference between the phase shifts in degrees at opposite azimuthal orientations of the tool. At line 1 the tool is entering the reservoir and the azimuthal resistivity in track 2 increases. The resistivity is somewhat amplified due to the polarization effect that occurs at boundaries with different resistivities, where electrons build up on the lower resistive boundary and protons on the high resistive boundary. This results in an oscillating charge across the boundary that amplifies the resistivity when moving across the formation and reservoir boundary [1]. The up and down resistivities in track 3 are at the same boundary. The geosignal in track 4 is positive since the tool is moving from a low resistivity formation to a high resistivity reservoir. As the tool moves towards the center of the reservoir, as shown between Lines 1 and 2, track 2 resistivity increases. Since the tool is now in the middle of the formation, both the up/down resistivities are the same value and therefore, the geosignal is zero. Between Lines 2 and 3 the tool approaches the lower resistive formation from the higher resistive formation and the geosignal shifts to negative since there is a decrease in the down resistivity as the tool approaches the low resistive formation and an increase in the up resistivity as it is detecting the high resistive formation. The tool is then corrected to navigate away from the low resistive formation and back into the center of the reservoir, as shown between Lines 3 and 4. The resistivity is higher at Line 4, and as shown between Lines 1 and 2 before, both the up/down resistivities are the same and geosignal is zero. As the tool approaches the top low resistive formation shown between Lines 4 and 5, the up resistivity decreases as the tool now is approaching a low resistive formation from the high resistive reservoir. The down resistivity is higher as it is facing the reservoir and therefore, the geosignal crosses back to the positive domain. Between Lines 5 and 6 the tool is steered again to the center of the reservoir, and as shown at Line 6, the geosignal is zero since the up/down resistivities are the same. Therefore, azimuthal resistivity measurements play an important role in steering the well within thin reservoirs surrounded by a resistive environment.

4.1.2 Triaxial Induction Tool

Recent advancements in computational-processing power have been utilized to build triaxial induction tools that obtain both horizontal and vertical resistivities of formations. Therefore, triaxial induction tools can be utilized to obtain resistivity measurements in dipping formations, where the bedding planes of the formations

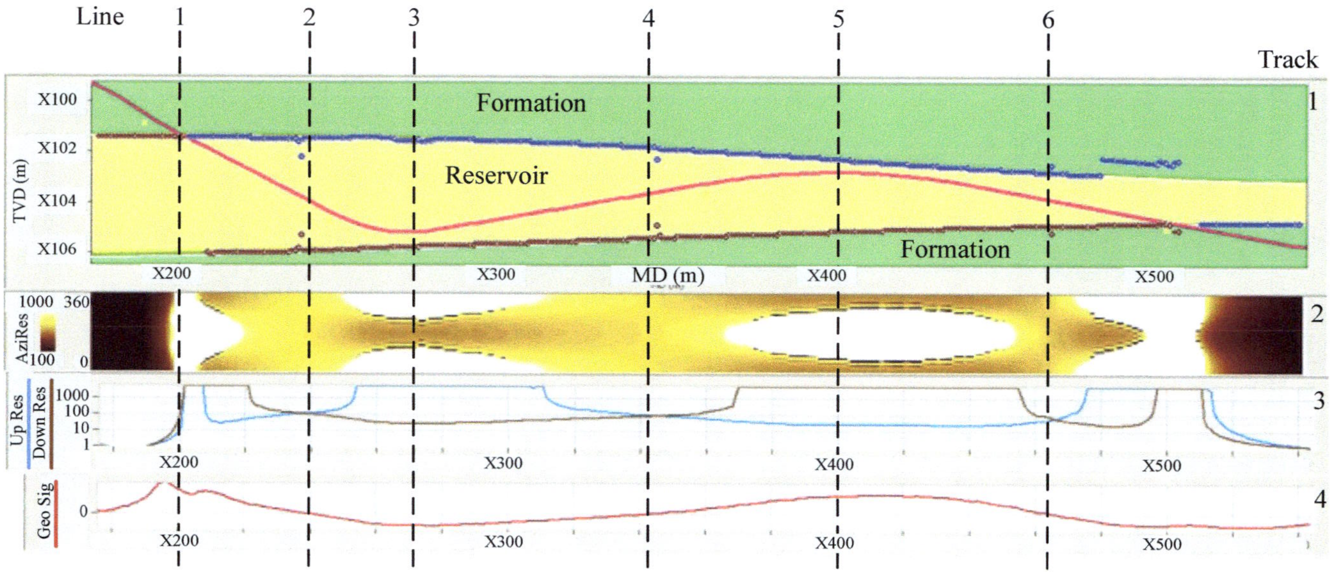

Fig. 4.4 Geosteering in a thin reservoir with the azimuthal propagation resistivity tool. Track 1 is a geological model of a high resistive thin reservoir sandwiched between a low resistive formation, track 2 is the azimuthal resistivity image, track 3 is the up and down resistivity and track 4 is the geosignal [1]

are not perpendicular to the tool, and identify and quantify laminated reservoir zones. The tool also has a new structural dip that does not require pad contact with the wellbore wall [5]. Laminated or thin reservoirs are challenging because the reservoir is thinner than the resolution of the tool and previous induction tools only measured the horizontal resistivity of a formation. Therefore, the properties of the reservoir were not fully resolved, which led to hydrocarbon zones often being overlooked or underestimated.

The triaxial induction tool is shown in Fig. 4.5a [5]. The tool consists of a triaxial transmitter, three short-spacing uniaxial receivers for wellbore corrections, and six triaxial receivers. Electrodes present on the tool and the mud resistivity (R_m) sensor at the nose of the tool at the bottom of the tool are also used for wellbore corrections. Wellbore corrections are needed to offset the influence variations

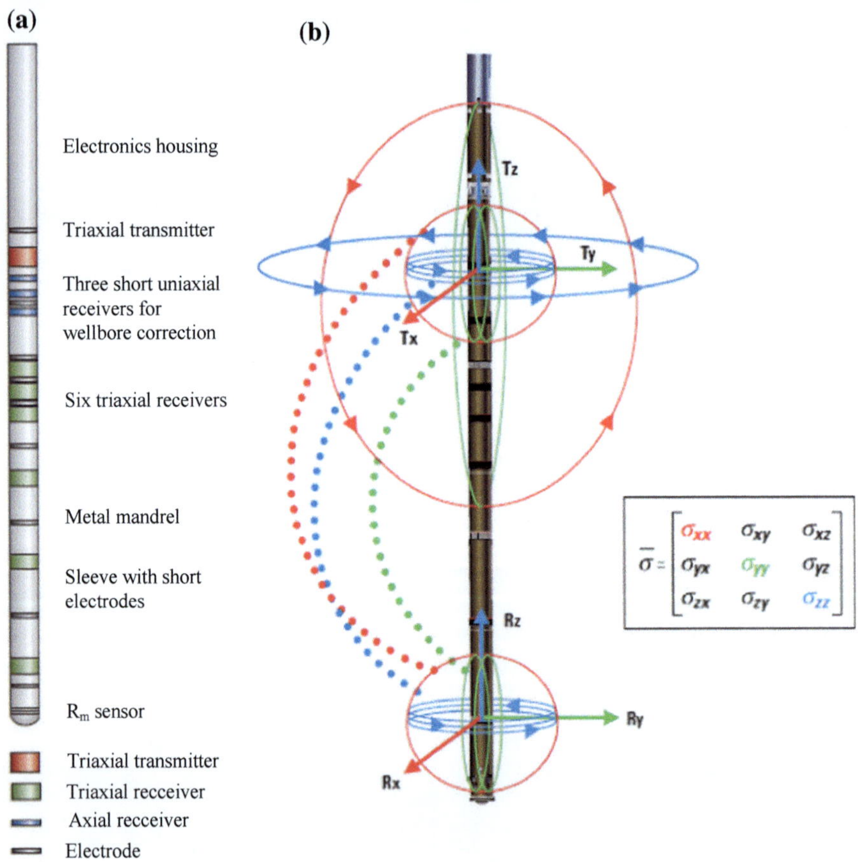

Fig. 4.5 **a** Triaxial induction tool with a triaxial transmitter, and six triaxial receivers. Three short uniaxial receivers, electrodes and a mud resistivity (R_m) sensor are used for borehole corrections. **b** Nine element, three-dimensional array produced by the triaxial tool, where the multiple triaxial transmitter and receiver pairs generate 234 conductivity measurements for each depth frame [5]

in size of the wellbore and/or tool eccentricity have on the measurements. A conductive path is provided by an internal mandrel (not shown in Fig. 4.5a) for wellbore currents to return through the electrodes on the exterior of the tool. Figure 4.5b shows the operation of the triaxial induction tool. While the traditional uniaxial induction tool has a single toroid shaped response in one direction, the triaxial tool shown in Fig. 4.5b produces nine measurements for each transmitter and receiver pairs, superimposed on each other [5]. While traditional induction tools have currents passing through coils wrapped around the tool, the z-axis, so that current is induced to flow in the formation concentrically around the tool, triaxial tools also have coils wrapped around the x-and y-axes. Therefore, triaxial tools also have current flowing in planes along the tool's x- and y-axes and the x, y and z components of the transmitter couple with the x, y and z receivers. In vertical wells only the xx, yy and zz components respond to the formation, but in deviated or dipping formations all nine components are utilized to resolve the resistivity measurement of the formation.

The triaxial tool measures both the horizontal resistivity (R_h) and vertical resistivity (R_v) and the ratio of R_v to R_h provides an indication of the level of anisotropy within a formation/reservoir, where a ratio higher than 5 indicates potential thin reservoirs. The couplings of the coils in the triaxial tool also provides several advantages when bedding planes of a formation/reservoir are not perpendicular to the relative position of the triaxial tool. For example, if the transmitter and receivers are not at the same position, the spacings for the cross-terms will be different from those of the direct terms. However, the measurements do not have to be depth-shifted to form the measurement tensors since all the measurements are performed within a single depth frame. Moreover, relative formation dip can be solved by mathematically rotating the nine component matrix since all the components are at the same spacing and location. A simple transformation can be performed when changing from one coordinate system to another since all measurements are obtained along the same coordinate system as well as at the same depth.

Figure 4.6 shows the log measurements obtained in a 60° deviated offshore well [5]. In track 1 are the caliper measurements indicating the size of the wellbore and natural gamma ray (GR) measurements. In track 2 is the depth of the well, track 3 are resistivity measurements from an uniaxial induction tool and track 4 are resistivity measurements from a triaxial tool. The encountered formations included two 30 ft formations with high resistivity, which was within the vertical resolution of the uniaxial tool. However, the deepest, 90-in array, induction resistivity measurements in track 3 were lower than the actual true resistivity values of the formation. The well has two zones of hydrocarbons indicated by high resistivity values. The uniaxial tool in track 3 measured 100 Ω m in the upper section, from X,940 to X,990, and 20 Ω m in the lower section (Y,000–Y,050). The triaxial tool measured much higher resistivity values, 500 Ω m, in the upper section and 100 Ω m in the lower section. These values by the triaxial tool corrected for the dip in the formation are five times higher than the deepest measurement obtained by the uniaxial tool. Moreover, in the lower 100 ft, Y,100 to Y,200, R_h values are

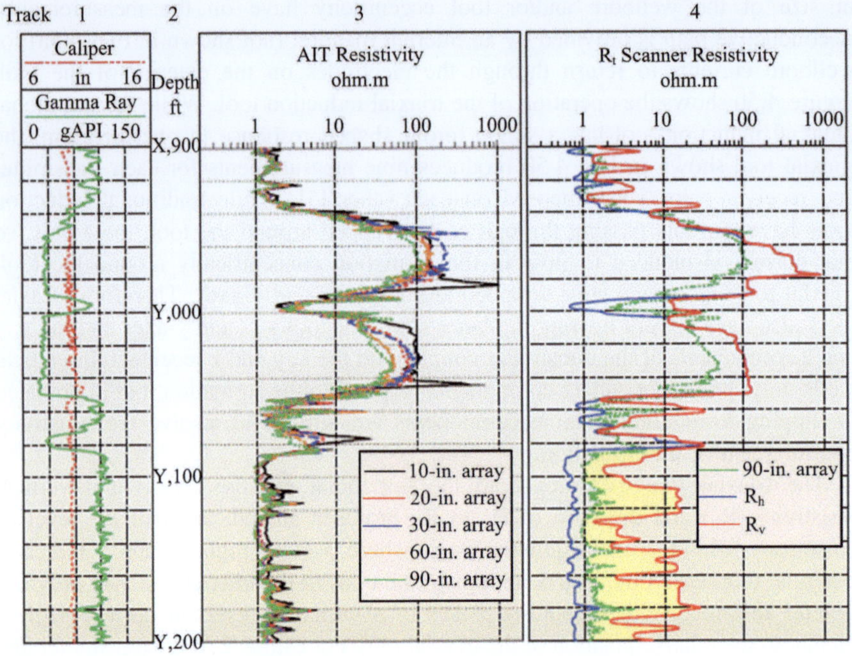

Fig. 4.6 Triaxial tool used in a well to accurately estimate resistivity of formations. Track 1 is the caliper and natural gamma ray measurements, track 2 is the depth, track 3 are measurements from a uniaxial induction tool and track 4 are the measurements from the triaxial tool. The triaxial tool vertical resistivity measurements in track 4 (red curve) provides a more accurate interpretation of hydrocarbon presence and volume compared to the uniaxial tool's deepest resistivity measurement (green curve) [5]

significantly lower than Rv values, which indicates anisotropy in the reservoir. Therefore, this anisotropy (in yellow shading) suggests a thin sequence of formations with additional hydrocarbon potential. While water saturation calculated utilizing the resistivity measurements obtained by both the uniaxial and triaxial tools will indicate the presence of hydrocarbons, the actual hydrocarbon reserves would be significantly different. This has far reaching implications since production facilities, long term infrastructure planning for future secondary and tertiary recovery programs depend on the volume of hydrocarbon reserves. Moreover, accurate estimation of hydrocarbon reserves would enable the petrophysicists and geologists to construct more extensive reservoir models with fewer wells. This is especially beneficial when exploring offshore fields, where the cost of drilling exploration wells to refine reservoir models is very high. Figure 4.7 shows an example where running a triaxial tool (track 3) showed much higher R_v values, more than 10 Ω m, than the uniaxial 90-in array values (track 2) in several intervals. Utilizing the resistivity values for the sand formation (note that lithology information is shown in track 5), water saturation, S_w, was calculated for both the uniaxial and triaxial measurements (track 4). The S_w values from the triaxial tool

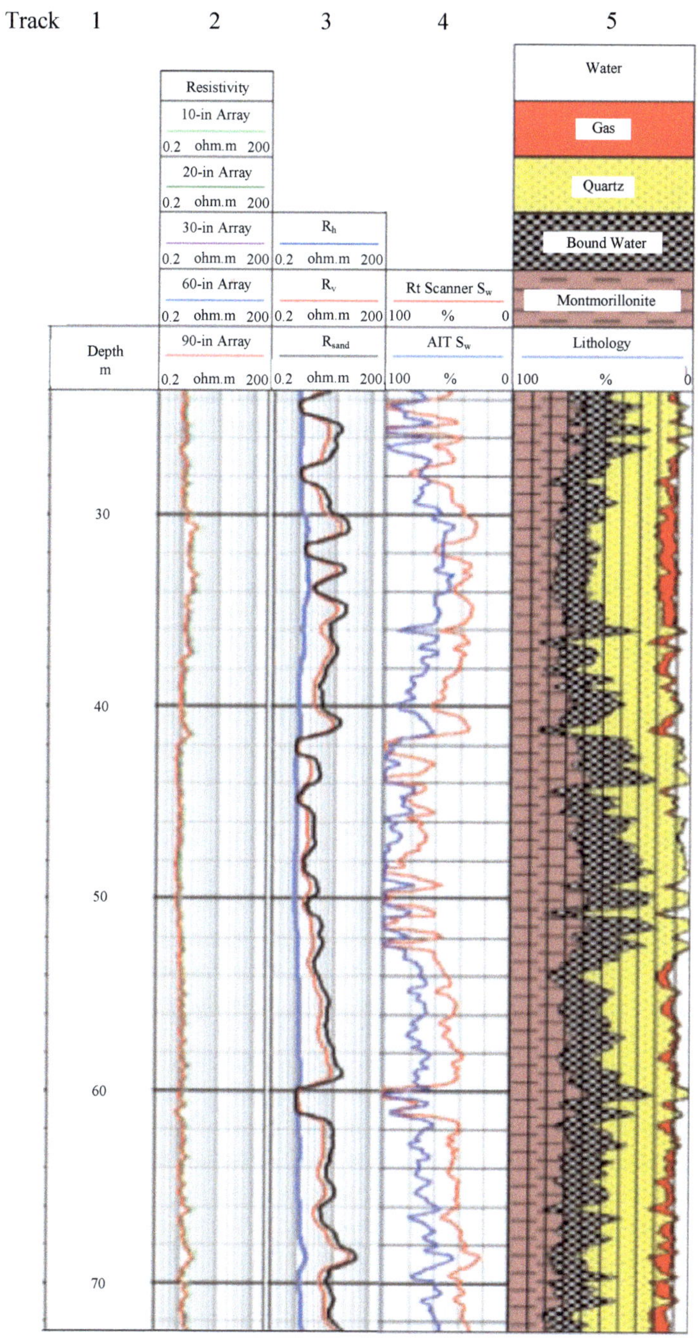

Fig. 4.7 Triaxial tool run in a well section showed much higher vertical resistivity values (track 3), than the uniaxial 90-in. array values (track 2), which led to the conclusion that there were more hydrocarbons in the reservoir than originally predicted from the uniaxial outputs [5]. Track 1 is depth, track 4 is water saturation calculated for both uniaxial and triaxial tools, and track 5 is the lithology of the rock formation [5]

outputs are significantly lower than the S_w from the uniaxial tool indicating the presence of more hydrocarbons in the reservoir than originally predicted from the uniaxial outputs.

4.1.3 Wireline Dielectric Tool

Early dielectric measurement were obtained at a single frequency of 1 GHz utilizing a fixed pad logging tool as shown in Fig. 4.8a. The fixed pad had two transmitters and receivers to obtain a wellbore compensated measurement. These initial tools were limited due to influences from wellbore rugosity and high salinity as well as the algorithms utilized to process the measurement data. The distance between the dielectric tool and the wall of the wellbore severely affected the quality of the measurements, and since high salinity of drilling fluids led to low signal-to-noise ratio the application of the tool was limited to fresh drilling fluids. The algorithms and processing techniques were also unable to accurately interpret the complex measurements. Figure 4.8b shows a modern dielectric tool with sensors on an extendable pad [6]. This design provides good contact with the wellbore wall thus reducing the effects of wellbore rugosity on the measurement data. The tool has two transmitters sandwiched between eight receivers, four symmetrically placed on either side of the transmitters as shown in Fig. 4.8c, which allows for wellbore compensation and pad tilt. The tool has the transmitters/receivers on the extendable pad and dielectric measurements of formations at DOIs of 1–4 in, at four separate frequencies between 20 MHz and 1 GHz. The tool also has improved algorithms and hardware that enables high quality data acquisition. The transmitters and receivers are able to simultaneously transmit and measure dielectric properties of formations in both longitudinal and transverse orientations. Figure 4.8d shows the response of the tool for different transmitter and receiver configurations. The signal response shows that measurements can be obtained directly in front of the center of the two transmitters. Moreover, the multi transmitter–receiver spacings allow various DOI while the upper and lower transmitter spacings provides high vertical resolution.

Figure 4.9 shows logging data from running the dielectric tool in a well [6]. Track 1 shows the residual oil volume (in shaded green) calculated from water saturation (SWXO), track 2 shows the depth of the well and track 3 shows standard resistivity measurements obtained from high resolution laterolog arrays and micro-cylindrically focused logs (RXO8, RLA5, RLA1, RT_HRLT) along with resistivity measurements derived from dielectric measurements (RXO_ADT). Track 4 shows porosity data obtained from the dielectric tool (PWXO_ADT), GR density tool (PHIT.PEQL) and a cross-plot of data obtained from GR and neutron porosity tools (PXND.PEQL). The shaded area in green is the area between PHIT.PEQL and PWXO_ADT and indicates residual hydrocarbon. The dielectric and density porosity values in track 4 are in good agreement in water-bearing zones while the dielectric resistivity and standard true resistivity values also match well. Residual

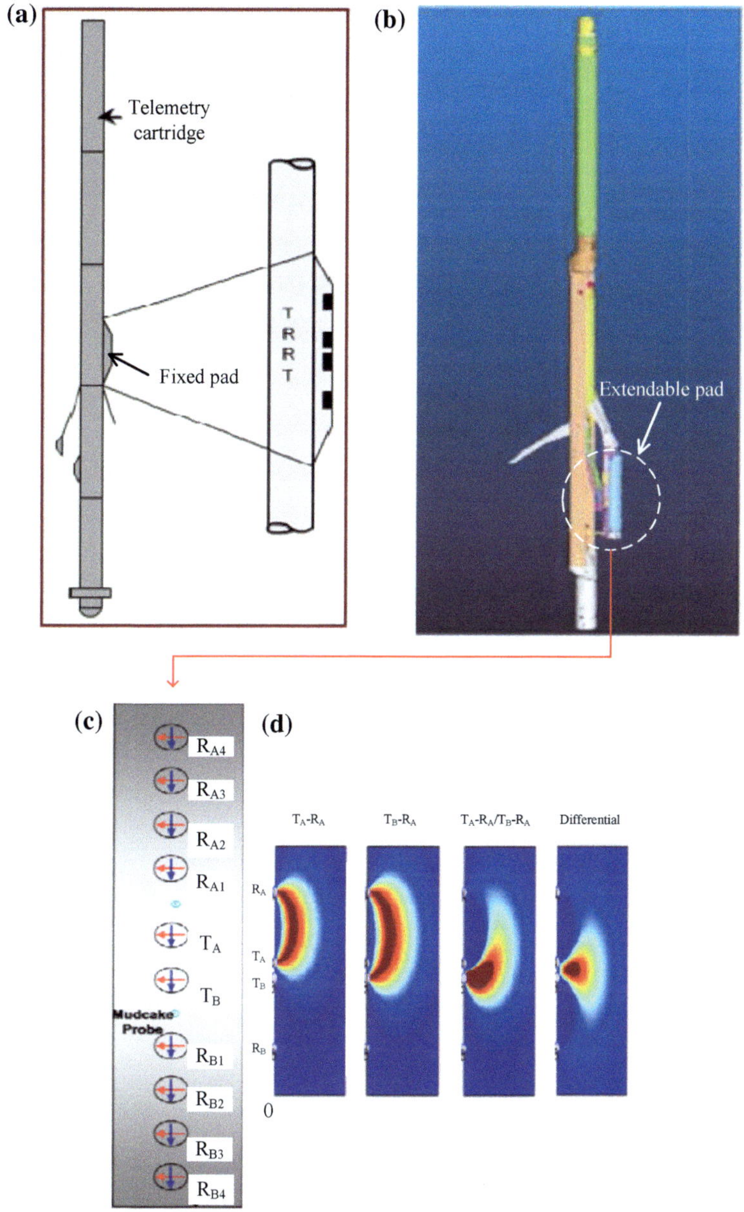

Fig. 4.8 Dielectric tool with **a** rigid pad, and **b** extendable pad with **c** transmitter and receiver layout on the pad and **d** response from the transmitter and receiver configuration [6]

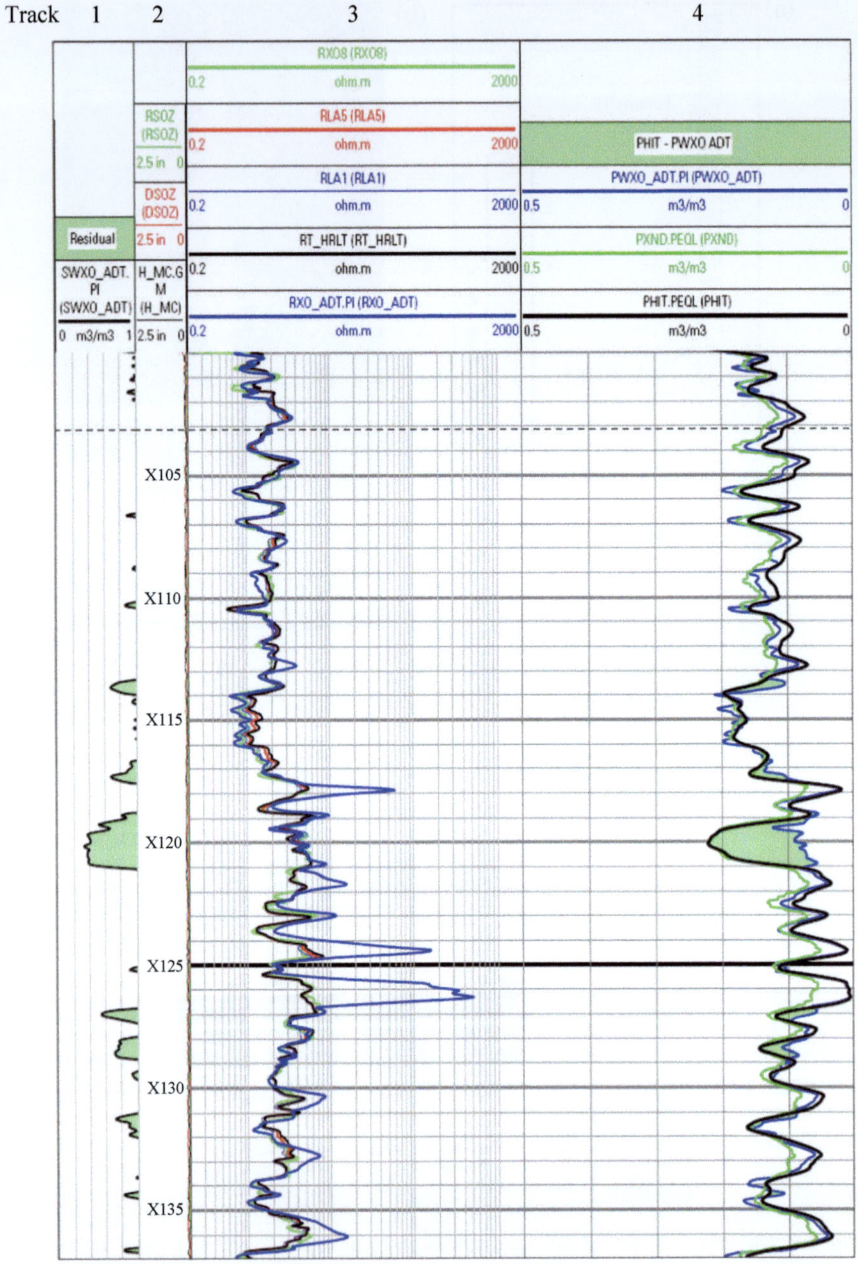

Fig. 4.9 Interpretation of a formation with the dielectric tool. Track 1 is the residual hydrocarbon calculated from water saturation values, track 2 is the depth, track 3 shows standard resistivity measurements obtained from high resolution laterolog arrays and micro-cylindrically focused logs (RXO8, RLA5, RLA1, RT_HRLT) and track 4 shows porosity data obtained from the dielectric tool (PWXO_ADT), a GR density tool (PHIT.PEQL) and a cross-plot of data obtained from GR and neutron porosity tools (PXND.PEQL). The shaded area in green indicates residual hydrocarbon [6]

hydrocarbons can be observed between X118.5 and X121 ft, where there is a difference between the dielectric porosity and density porosity values and also an increase in the resistivity values.

4.1.4 *Spectral Azimuthal Gamma Ray Tool*

Spectral GR measurements provide information on not only the total GR but also the potassium (K), thorium (Th) and uranium (U) content in formations. Spectral measurements are useful when evaluating advanced clay typing and qualitative measurement of Total Organic Carbon (TOC). The TOC is strongly correlated to U and advanced clay typing indicates the brittleness or ductility of unconventional shale formations. These shale formations have low K and Th concentrations and very high U concentrations, and natural or induced fracture networks in these formations result in significant hydrocarbon production potential. Standard MWD/LWD GR tools measure the total gamma radiation in formations but are unable to identify K, Th and U from total radiation measurements. Wireline GR tools are able to identify and separate K, Th and U from total GR data but measurements cannot be acquired in real-time, while drilling.

An LWD spectral azimuthal GR (SAGR) tool with sodium iodide (NaI) detectors were utilized to obtain spectral measurements in a well. The SAGR tool differs significantly from standard GR tools since the detectors are mounted on the outside of the tool as shown in Fig. 4.10a [7]. Standard GR tools can be either insert-based tools, as shown in Fig. 4.10b or sonde-based tools, as shown in Fig. 4.10c. Insert-based tools are normally used on a LWD platform, where the detectors placed in an insert are mounted into a drill collar as shown in Fig. 4.10b. In a downhole environment this configuration requires the GRs to travel through the drilling fluid in the wellbore and the drill collar. Moreover, the type of drilling fluid

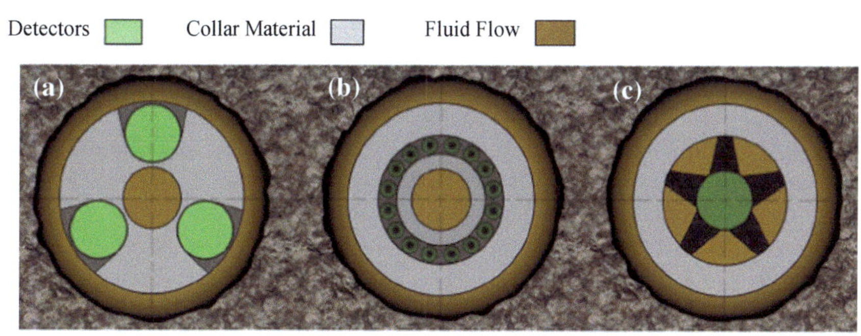

Fig. 4.10 Gamma ray tool configurations. **a** Spectral azimuthal gamma ray (SAGR). **b** Insert-based gamma ray. **c** Sonde-based gamma ray [7]

and detector can result in a significant amount of GRs being absorbed by the collar and drilling fluids such as barite, even before the GRs reach the gamma detectors. While this type of configuration can be utilized for azimuthal binning, the image quality is generally lower due to the low count rates at the detector, especially if the rate of penetration during drilling is high. Sonde-based tools are almost exclusively utilized on MWD platforms and have the gamma detector in a pressure housing located at the center of the collar as shown in Fig. 4.10c. Similar to the insert-based tools the GRs in sonde-based tools have to travel from the formation, through the drilling fluid and the collar to the detectors. The drilling fluid and drill collar attenuation result in poor statistical accuracy and measurements being limited to total gamma ray measurements making it challenging to build a tool that can obtain azimuthal measurements. The SAGR tool on the other hand has the detectors on the outside of the drill collar, which allows a maximum number of GRs to travel through a path from the formation, through the drilling fluid to the detectors. Apart from the increase in the number of GRs reaching the detector, there is also an increase in the precision of the measurements due to less attenuation. The result is a more accurate gamma log, sharper images, better formation dip determination and obtaining results at significantly higher drilling speeds.

Figure 4.11 shows the comparison between the SAGR tool and an insert-based tool in a horizontal well [7]. Track 1 shows the rate of penetration of the drilling assembly in ft/h, wellbore drift in degrees and total GRs detected by the SAGR tool with a significant peak at X100 and minor peaks at X180, X285, X440, X480, X540, from X720 to X860, at X920, X980 and X1040. The high-temperature azimuthal gamma ray (HAGR) static image in track 7 were obtained from a tool that contains a ring of Geiger-Müller tube detectors, while the static image in track 4 were obtained from the SAGR tool. Static images utilize a fixed scaling range while dynamic images adjust and optimize the scaling ranges according to pre-determined intervals. Comparing the HAGR image in track 7 and the SAGR image in track 4, it can be clearly seen that the SAGR image is of superior quality to the HAGR image. Moreover, the dynamic image obtained by the SAGR tool in track 6 is also of high quality, where the boundaries outlined in green in track 4 can be clearly identified in track 6. Another contribution to the difference in the quality between the SAGR and HAGR images is that the HAGR tool has an 8-bin image, which is not as azimuthally focused as the SAGR tool due to collar shielding. Furthermore, the SAGR does not have spectral biasing and has a more precise vertical resolution than the HAGR tool due to the configuration of the detectors in the tool. The quality and resolution of a GR image plays a key role in formation bed analysis and dip picking interpretation, which in turn aids to optimize geosteering in real-time. One of the most important features of the SAGR tool is the accurate measurement of U, Th and K in formations, which helps to evaluate organic richness and clay content in formations. Track 3 shows the characterization of these elements and how peaks in U can be clearly identified at the total GR peaks in track 1.

Figure 4.12 shows an SAGR tool that was run inside a well with a standard azimuthal density (AZD) tool [7]. There is an increase in the total GR measurements in track 1 from X200 to X1000 and this is clearly reflected in both the static

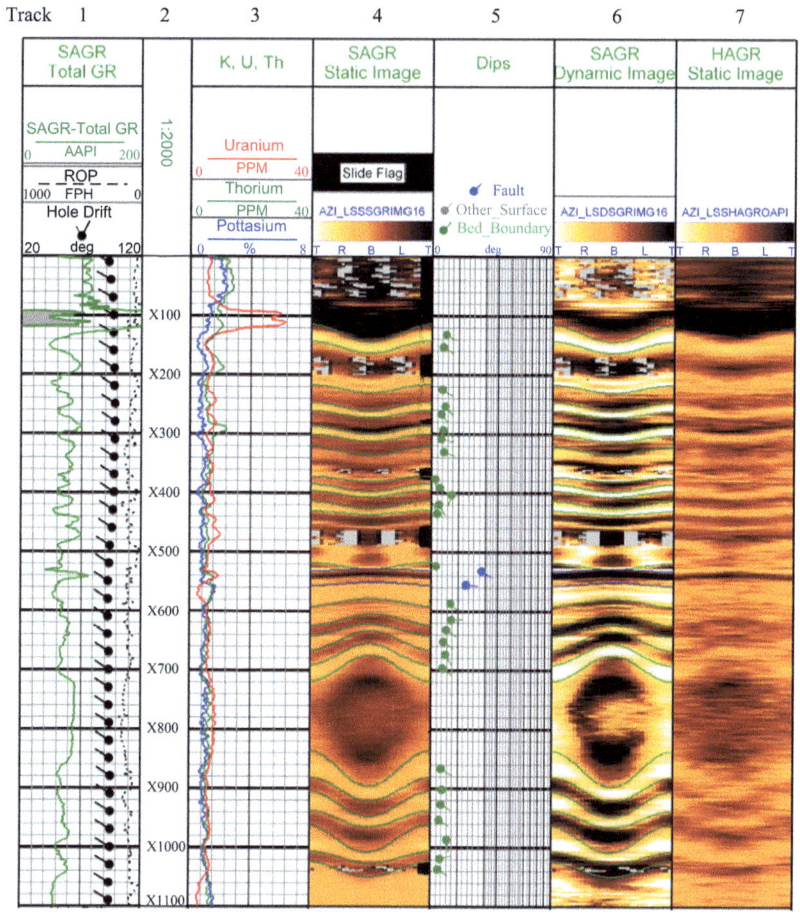

Fig. 4.11 Comparison between the measurements acquired in a well by the spectral azimuthal gamma ray (SAGR) tool and an insert-based high-temperature azimuthal gamma ray (HAGR) tool. Track 1 shows the rate of penetration of the drilling assembly in ft/h, wellbore drift in degrees and total GRs detected by the SAGR tool. Track 2 is the depth, track 3 is the elemental composition, track 4 is the static image obtained by the SAGR tool, track 5 are the dip measurements, track 6 is the dynamic image obtained by the SAGR tool and track 7 is the static image obtained by the HAGR tool [7]

and dynamic images obtained by the SAGR and AZD tools in tracks 5 and 7–9. It is interesting to note that the density measurements in track 4 show higher values from X200 to X700, approximately 2.4 g/cc, compared to values from X700 to X1100, approximately 2.3 g/cc. The difference can be explained by the spectral measurements in track 3, which show higher lower U values for from X200 to X700 and higher values from X700 to X1100. This indicates the presence of high organic content from X700 to X1100 since high U content is directly proportional to low density and therefore, high porosity.

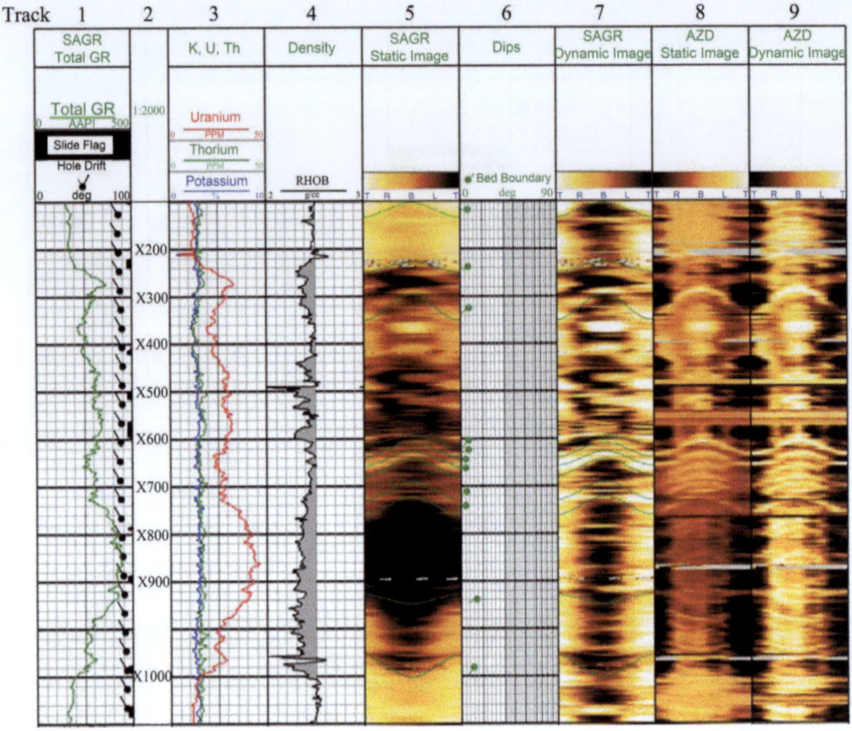

Fig. 4.12 Comparison between the measurements acquired in a well by the spectral azimuthal gamma ray (SAGR) tool and an azimuthal standard density (AZD) tool. Track 1 shows the total GRs detected by the SAGR tool and wellbore drift in degrees. Track 2 is the depth, track 3 is the elemental composition, track 4 is the density measurements by the AZD tool, track 5 is the static image obtained by the SAGR tool, track 6 are the dip measurements, track 7 is the dynamic image obtained by the SAGR tool, track 8 is the static image obtained by the AZD tool and track 9 is the dynamic image obtained by the AZD tool [7]

4.1.5 Pulsed Neutron Tool

Pulsed neutron (PN) tools eliminate radioisotope sources utilized in conventional gamma and neutron porosity tools. While radioisotope sources have been utilized for obtaining density and porosity of formations for decades, there are many restrictions and safety issues surrounding the use of these tools in wells. Radioisotope based tools are enclosed in special housings that prevent rig personnel from exposure and are kept in isolation prior to use in wells. When deployed inside wells the tools are housed in pressure vessels that protect the radioisotope sources from mechanical vibrations, shocks, temperature and corrosion in the harsh downhole environment. There are also health and safety guidelines and exposure limits established to protect rig workers who deploy these tools. The PN tool presented in this section employs a pulsed neutron generator instead of a radioactive

PNG-Based Measurements	Other Measurements	
Neutron-gamma density	Array resistivity	Azimuthal gamma ray
Neutron porosity	Dual ultrasonic caliper	Near-bit inclination
Spectroscopy	Annular pressure while drilling	Three-axis shock and vibration
Sigma	Temperature	

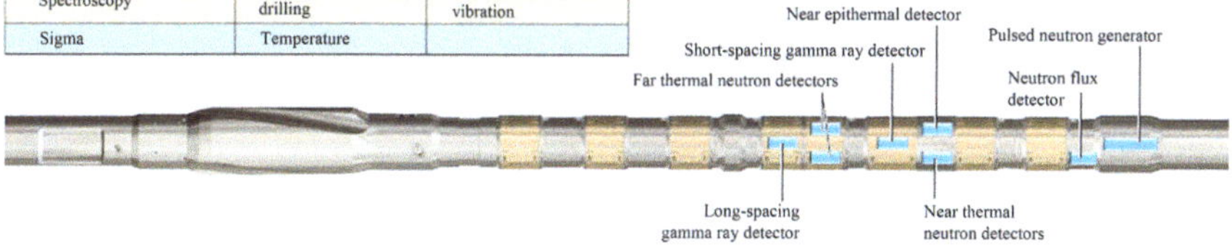

Fig. 4.13 The pulsed neutron (PN) tool [8]

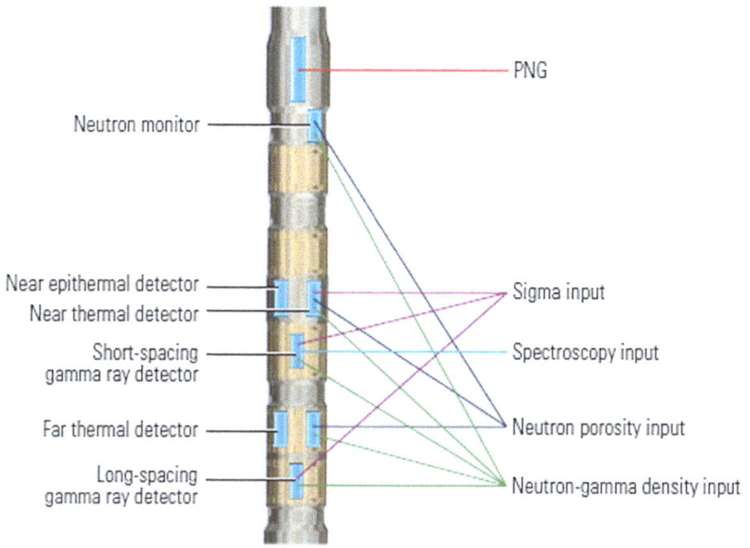

Fig. 4.14 Multiple measurements that can be acquired by the pulsed neutron tool [8]

source to generate neutrons, and can obtain density, thermal neutron porosity, elemental capture spectroscopy, capture cross-section sigma, 2 MHz and 400 kHz propagation resistivity and azimuthal natural GR measurements. The PN tool is shown in Fig. 4.13 and it is located in the vicinity of the drill bit, thus, providing real-time measurement data for drilling engineers to accurately drill a well [8]. Figure 4.14 shows how the responses from multiple detectors are integrated to produce specific measurements. For example, once the pulsed neutron generator (PNG) generates neutrons the density is primarily derived using counts from the long-spacing GR detector but also takes into account inputs from the neutron monitor, near epithermal detector, short- and long-spacing GR detectors and far thermal detectors. The ability to measure formation density without a traditional Caesium (Cs-137) source is one of the most important features of this tool. Formation density is a critical parameter in the determination of formation porosity.

While radioisotope sources were replaced in neutron porosity tools before the Cs-137 have only being replaced recently in density tools. Several reasons, such as priority given to replacing radioisotopes in neutron porosity tools since the sources had a much longer half-life, the unavailability of a comparable electronic GR generator and replacing a straightforward density interpretation accepted by the industry with a more complex method, delayed replacing sources in density tools with PNGs.

The PN tool has a miniature PNG that produces high energy, fast neutrons using a charged particle accelerator. As shown in Fig. 4.15, ionized deuterium (^{2}H) and tritium (^{3}H) isotopes of hydrogen are accelerated towards a tritium-doped target by a high voltage potential, resulting in a fusion reaction producing a ^{4}He nucleus and a neutron [8]. When the particles are stopped in matter the energy from the fusion reaction is converted to kinetic energy of the two particles. The high energy neutrons are released from the reaction with very high speed, having kinetic energy of 14 meV of the total 17.6 meV released. Unlike radioisotope based sources, the PNG does not emit radiation once the main power is disconnected.

Figure 4.16 shows the life of the neutrons released in short bursts by the PNG [8]. These neutrons interact with the atoms and molecules of the formation rocks and fluids, where most of these interactions are elastic collisions with hydrogen nuclei in the formation. After successive collisions the high energy neutrons slow down and reach thermal energy. Traditional neutron porosity tools count these thermal neutrons that arrive back at the tool, which is used to derive the formation porosity. However, not all the collisions are elastic and during the first few microseconds, before reaching the thermal energy level, neutrons experience inelastic collisions with atomic nuclei in the formation. These inelastic collisions elevate some neutrons to an excited state and result in the emission of one or more GRs as they return to the ground state. Elements such as carbon, oxygen, silicon,

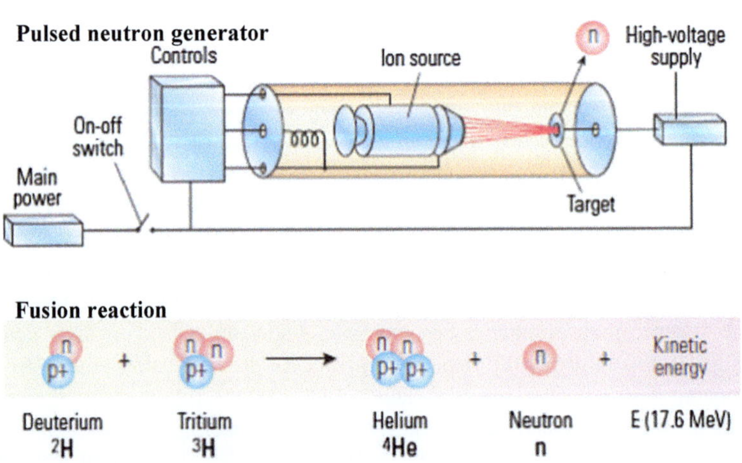

Fig. 4.15 A pulsed neutron generator (PNG) that generates neutrons by a fusion reaction when the main power is connected [8]

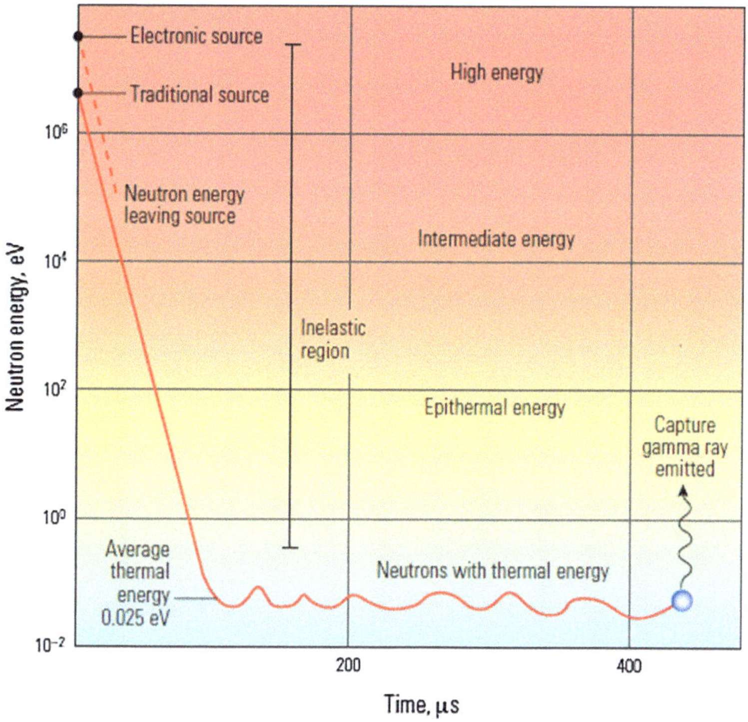

Fig. 4.16 The life of a neutron after it is released in short bursts by the pulsed neutron generator (PNG) [8]

calcium, iron and sulfur can be identified by using the energy spectrum of the inelastic GRs and the volumetric quantification of these elements can be used to characterize lithology, which is known as neutron spectroscopy.

A GR cloud with around 108 GR/s is formed momentarily during the inelastic collisions of the neutrons with the atomic nuclei. These GRs travel back to the tool and are counted by the detectors, where the number of GRs depend on the electron density of the formation. As the GRs move through the formation they collide with electrons in the formation primarily through Compton scattering and are attenuated. The counts at the detector are characterized and correlated to the density of the formation. However, fast neutrons reach the thermal energy level and are captured by atoms in the formation. GRs are emitted by the nuclei that capture the thermal neutrons and the population density of thermal neutrons available for capture is dependent on the number of hydrogen atoms in the formation. Therefore, when computing bulk density the contribution from the GRs resulting from neutron capture is removed from the density measurement.

Figure 4.17 shows the measurements obtained by a PN tool in a well with an average inclination of 60° [8]. Track 1 shows the deviation of the well in degrees, the rotation of the tool in RPM and the depth of the well. There is a hydrocarbon

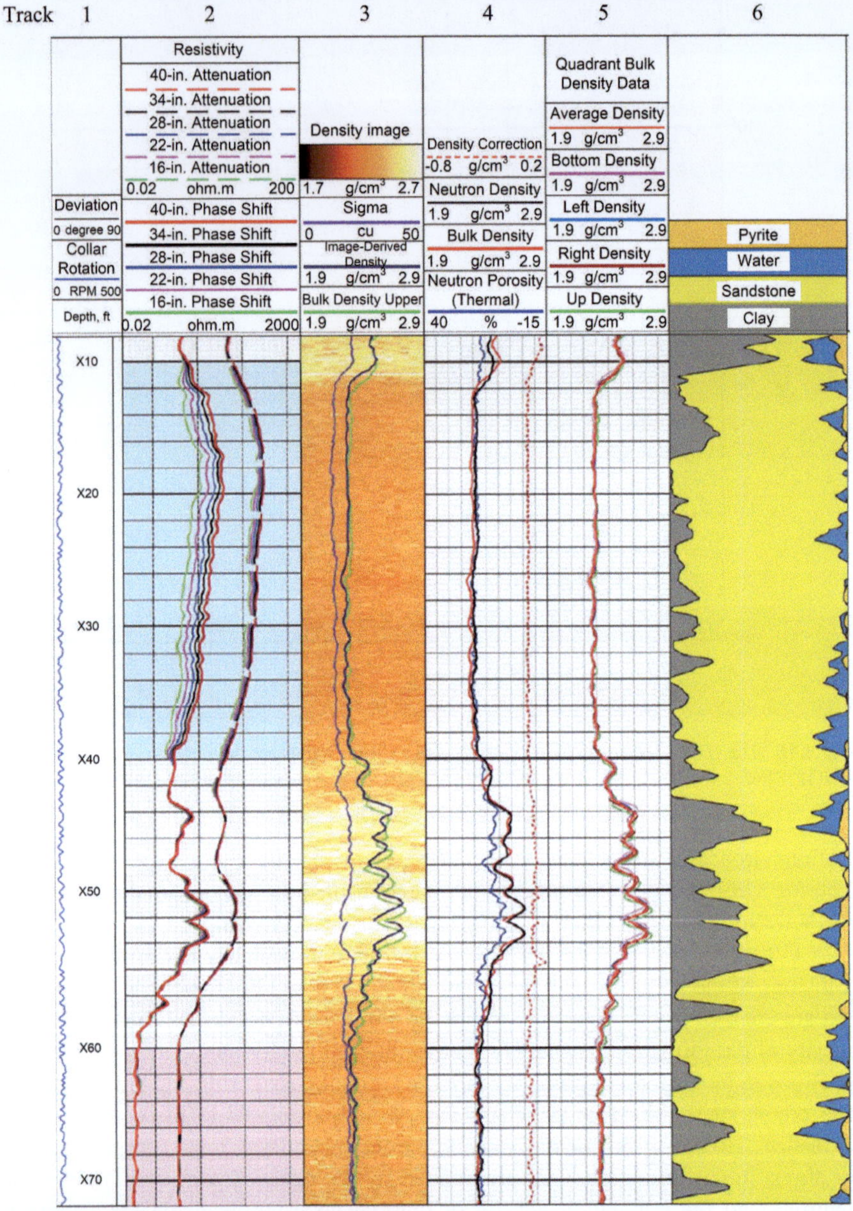

Fig. 4.17 Comparison of the PN tool with a standard gamma ray tool to measure density in formations. Track 1 shows the deviation of the well in degrees, the rotation of the tool in RPM and the depth of the well, track 2 shows standard resistivity measurements and tracks 3 and 5 show the density image obtained by a standard gamma ray density tool. In track 4 neutron density measurements obtained by the pulsed neutron (PN) tool is compared with the bulk density measurements acquired by the gamma ray density tool. Porosity measurements by a neutron porosity tool is also included in track 4 and track 6 shows the lithology of the formation [8]

zone in the sandstone formation (track 6) from X10 to X40 ft (blue shaded area in track 2) as indicated by the low bulk density measurements from a standard gamma ray tool in tracks 3–5 and neutron density measurements by the PN tool in track 4. There is excellent agreement between the standard gamma ray tool measurements (track 4, red line) and PN tool neutron density measurements (track 4, black line). Moreover, the density image in track 3 indicates a homogenous reservoir and this is further clarified by the bulk density measurements overlying each other in track 5. The increased resistivity measurements in track 2 support the presence of hydrocarbons and the separation of the resistivity curves indicates the diameter of invasion of the wellbore by drilling fluids. The zone from X60 to X70 in the formation (red shaded area in track 2) is water-bearing sand, since even though density measurements (tracks 3–5) and porosity measurements (track 4) are similar to the X10–X40 zone, the resistivity measurements in track 2 are lower. The sigma measurement in capture units (c.u) in track 3 refer to the macroscopic thermal neutron capture cross section of the formation. In PN tool measurements, the rate at which thermal neutrons are captured is dependent on sigma. Since sigma measurements in track 3 are lower than 20 c.u., it can be concluded that sigma correction on the PN tool is minimal [8, 9].

4.1.6 Nuclear Magnetic Resonance Tool

In recent times there has been a steady increase in drilling more deeper and complex reservoirs. Most of these deeper reservoirs require slimhole drilling, where the size of the wellbore is generally less than 8–1/2 in. Comprehensive logging evaluation plays a critical role in slimhole drilling applications to optimize drilling and successfully reach the target reservoir. Nuclear magnetic resonance (NMR) tools not only provide information about formation porosity without a radioisotope source but also offer insight into more complex tasks such as identification of heavy oil/tar zones and hydrocarbon typing. The design of an NMR tool with an outer diameter of 4–3/4 inches utilized for slimhole drilling is shown in Fig. 4.18 [10]. The tool has two cylindrical permanent magnets positioned to face each other with like poles and magnetized in the axial direction to create a low gradient magnetic field (B_0). The axis symmetry of the magnets makes the NMR measurements rotation independent and also minimizes magnetic interference to other components in the tool due to the magnetic fields of the two magnets cancelling at far distances from the tool. Since a low lateral static gradient makes the MR tolerant against lateral tool vibration and the reduction in sensitivity of the NMR volume in the axial direction is not an issue while drilling, the NMR sensor was optimized for a low lateral static field gradient rather than an axially extended sensitive MR volume. The RF magnetic field pulses (B_1) that act on the nuclear spins in the formation is transmitted by the transmitter antenna and the echo signals from the formation is detected by the receiver. Both the transmitter and receiver antennas have a soft magnetic core underneath to enhance efficiency. The slimhole NMR LWD tool on the field is shown in Fig. 4.19 [10].

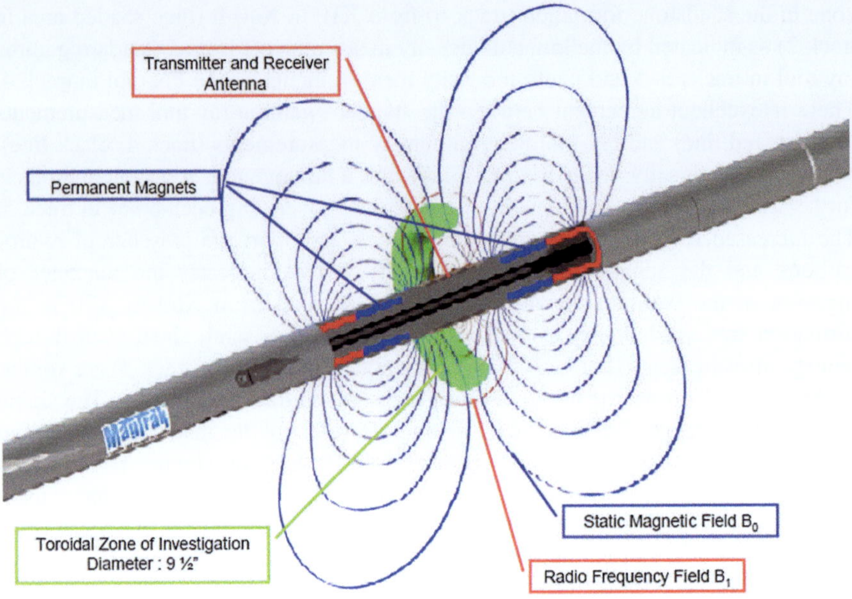

Fig. 4.18 The slimhole nuclear magnetic resonance (NMR) tool with two cylindrical magnets that produce a static magnetic field, a transmitter to create a radio-frequency (RF) magnetic field and a receiver to acquire signals from the formations [10]

Figure 4.20 shows a comparison between the NMR tool and a standard neutron-density tool [11]. Track 1 shows the resistivity of the formation, where the separation between the shallow and deep logs gives an indication of the diameter of invasion by the drilling fluid and the permeability of the different zones in a formation. Track 2 shows the lithology of the formations, such as anhydrite, dolomite, calcite and shale formations and Track 3 shows the depth at which measurements were made. Track 4 shows the porosity measurements, where the red curve shows the total porosity measurements obtained by the neutron porosity tool (TP-ND) and the yellow and gray shaded areas show the total porosity and bound fluid volume, respectively, obtained by the LWD NMR tool (TP-NMR and bound fluid volume (BFV)). Track 5 shows the T_{II} distributions of the NMR measurements. The images in Track 4 show that there is excellent agreement between the NMR and neutron-density results when computing total porosity results since the porosity percentages of the red curves and the yellow shaded areas are similar throughout the depths at which the logs were taken.

Figure 4.21 shows another comparison between a neutron-density tool and an NMR tool [11]. Track 1 is the resistivity, 2 the lithology and 3 the depth, as in Fig. 4.20. However, the NMR tool was run twice in the well since the measurements obtained during the first run, NMR-1 in Track 5, did not compare well with measurements obtained by the neutron density tool in Track 4. The neutron density tool measurements are shown in track 4, where the blue-shaded area corresponds to

Fig. 4.19 The fabricated slimhole nuclear magnetic resonance (NMR) tool being lowered into a well [10]

volume of water and the red-shaded area to hydrocarbons. The measurements obtained by the NMR tool during its initial run (NMR-1) is shown in track 5, where the yellow shaded area is the total porosity (TP-NMR) and the grey shaded area is the bound fluid volume (BFV). The red curve seen in this Track corresponds to the outline of the total porosity obtained by the neutron density tool (TP-ND) in Track

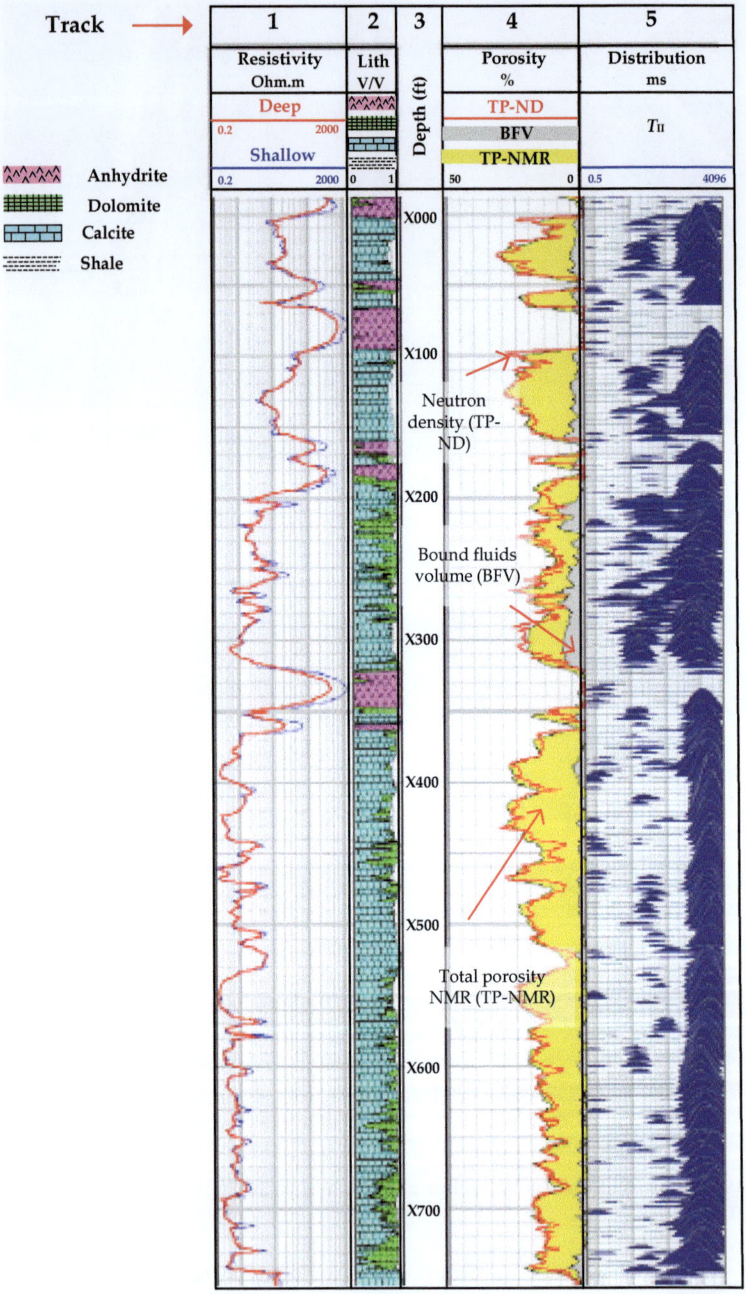

Fig. 4.20 Measurements obtained by the slimhole nuclear magnetic resonance (NMR) tool compared with measurements obtained by a standard neutron density tool. Track 1 is resistivity measurements by a standard resistivity tool, track 2 is the depth, track 3 is the lithology of the formation and track 4 shows the porosity measurements obtained by NMR and neutron density tools and track 5 shows the T_{II} distributions of the NMR measurements [11]

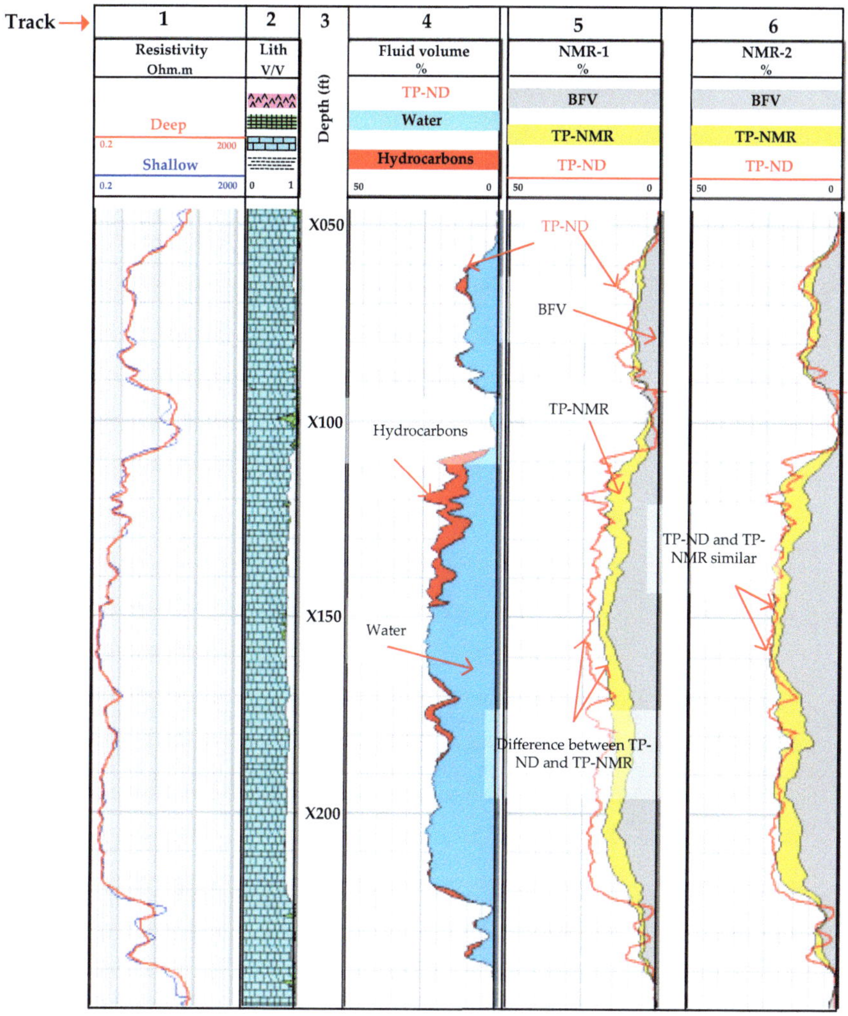

Fig. 4.21 Comparison between the slimhole nuclear magnetic resonance (NMR) tool and a neutron-density tool. Track 1 is the resistivity, track 2 the lithology and track 3 is the depth. Track 4 shows the measurements obtained by the neutron-density tool, where the blue-shaded area corresponds to volume of water and the red-shaded area to hydrocarbons. The NMR tool was run twice in the well and Track 5 (NMR-1) shows the measurements obtained during its initial run. The yellow shaded area is the total porosity (TP-NMR) and the grey shaded area is the bound fluid volume (BFV). The red curve seen in this Track corresponds to the outline of the total porosity obtained by the neutron density tool (TP-ND) in Track 4, and is added as a reference to Track 5 for comparison with TP-NMR. Track 6 (NMR-2) shows the measurements obtained during the second run of the NMT too [11]

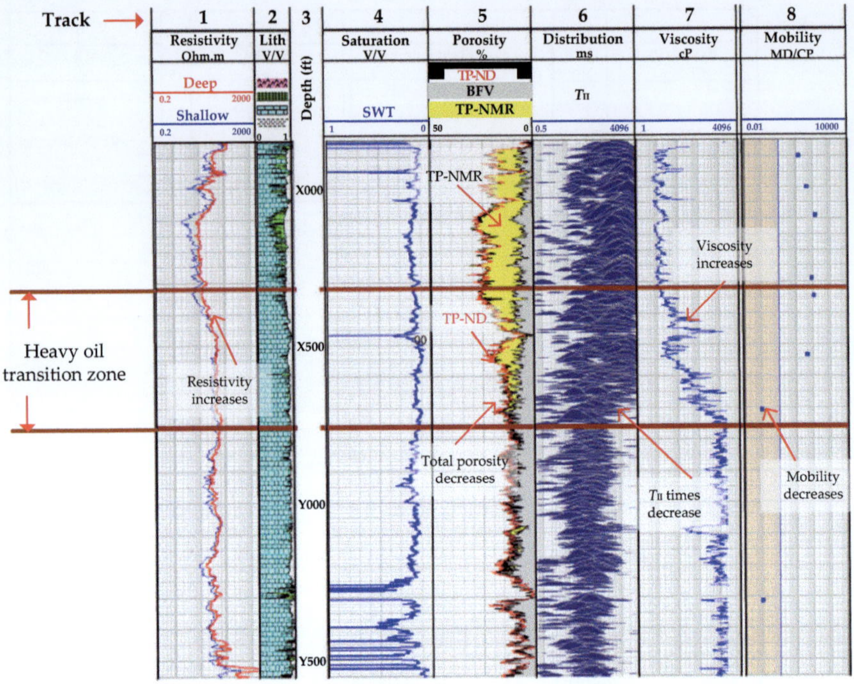

Fig. 4.22 Utilizing the slimhole nuclear magnetic resonance (NMR) tool to identify a heavy oil transition zone. Track 1 is the resistivity, track 2 is the lithology, track 3 is the depth, track 4 is the water saturation, track 5 is the porosity obtained by the NMR tool and a neutron-density tool, track 6 shows the T_{II} distributions of the NMR measurements, track 7 are the viscosity measurements and track 8 mobility measurements [11]

4, and is added as a reference to Track 5 for comparison with TP-NMR. The clear difference between TP-ND and TP-NMR was attributed to the effect of high salinity drilling fluid used inside the well. During the second run shown in track 6 (NMR-2) the tool was adjusted to compensate for high salinity, which resulted in the TP-ND and TP-NMR to be in good agreement.

Figure 4.22 shows how a heavy oil transition zone in the reservoir was identified by estimating the viscosity [11]. The heavy oil transition zone shown between the two brown lines can be identified by the decrease in the total porosity measurements obtained by both NMR (TP-NMR) and neutron density (TP-ND) in Track 5 due to pores being saturated with heavy oil, the increase in viscosity in track 7, the decrease in the mobility in Track 8, which is the ratio of effective permeability to phase viscosity and the T_{II} distribution times becoming shorter due to restricted molecular motion. A less obvious indicator is resistance in Track 1, which increases slightly in the heavy oil transition zone.

4.1.7 Acoustic Tool

Sonic logging tools acquire acoustic data in downhole environments, which along with high temperature and pressure also has significant noise and vibration. The propagation of acoustic waves through formations aid in the determination of pore pressure, overburden gradients, lithology and geomechanical properties, including fracture evaluation. Early LWD sonic tools only primarily provided compressional measurements with shear measurements in only some formations. The LWD sonic tool presented in this section uses quadruple sonic measurements to extract shear data in soft formations and, along with compressional data, can be used to monitor real-time pore pressure while drilling, improve well completions and estimate geomechanical formation properties.

The design of the LWD sonic tool that acquires both monopole and quadrupole measurements is shown in Fig. 4.23a and the fabricated tool is shown in Fig. 4.23b [12, 13]. It has a wideband multipole transmitter and 48 wideband receivers. The wideband multipole transmitter excites the wellbore monopole and quadrupole modes over a frequency range from 1 to 20 kHz. There are 4 receiver stations with 12 axial receivers, 4 in. apart, at each receiver station. The receiver stations are positioned 90° apart in protected grooves on the outside of the tool. In the monopole mode sound waves are transmitted by the transmitter in a radial pattern and travel through the formation as a compressional wave, as shown in Fig. 4.24a, to obtain the compressional slowness of the formation [13]. The slowness of a formation computed at the receivers of the sonic tool refers to the travel time of sound waves from the transmitter, through the drilling fluid, along the formation, back through the drilling fluid and finally to the receivers. Table 4.1 shows the compressional and shear wave mode slowness times, in μs/ft, for different downhole formation materials and liquids [13]. Generally rock formations with water, oil or gas in the rock's pore space have longer travel times than through rock formations with no porosity such as sandstone and limestone. Monopole sources can measure shear slowness only if the shear slowness of the formation is less than the compressional slowness in the wellbore fluid, a situation known as a 'fast' formation. This limitation of the monopole source, to only obtain shear measurements in fast formations, is overcome by quadrupole sources. The sound waves transmitted in the quadrupole mode, as shown in Fig. 4.24b, have complex waveforms that are frequency dependent [13]. At low frequencies they travel at velocities comparable to that of shear waves and therefore, can be utilized to determine shear in slow formations. The blue star in Fig. 4.24b refers to the approximate location along the wellbore of the wave represented in the cross-section.

Acoustic data can be used to obtain pore pressures of formations, derive elastic properties and tie downhole data with seismic data obtained at the surface. Pore pressure is a critical parameter while drilling as this aids drilling engineers to optimize drilling programs and maintain well stability by adjusting the density of drilling fluids to balance the formation pressures in real-time. One application of the sonic tool is the identification of overpressured formations, where optimizing the

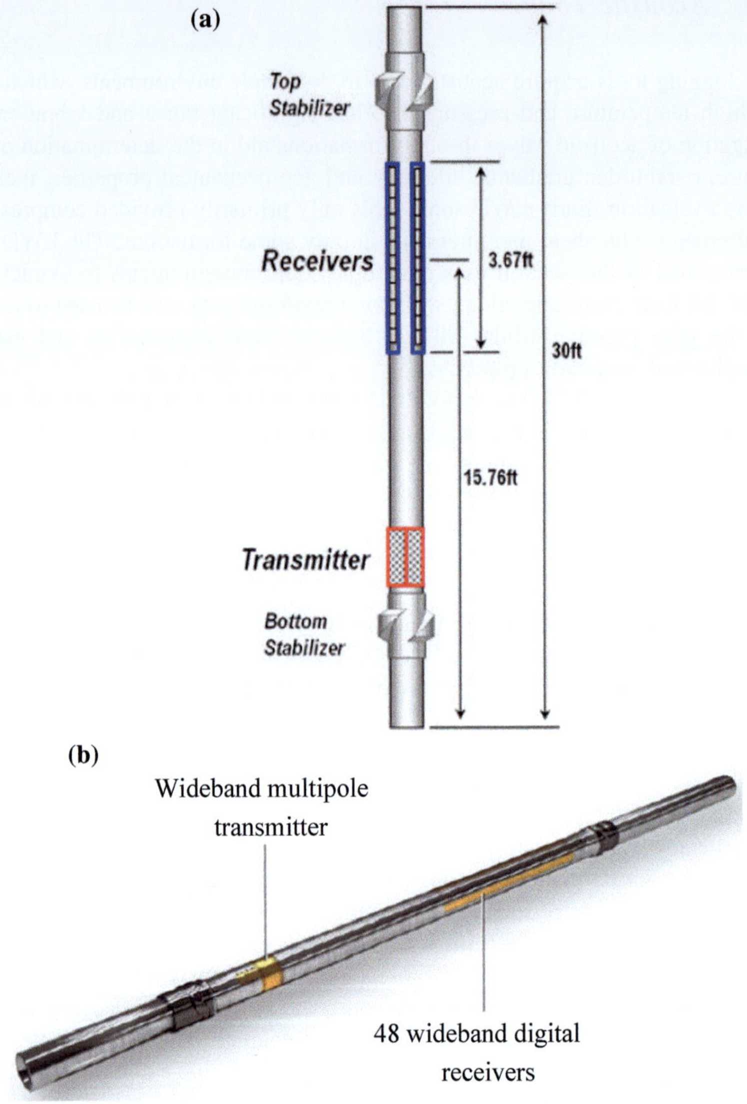

Fig. 4.23 **a** Design of the LWD sonic tool. **b** Fabricated LWD sonic tool [12, 13]

mud weight is imperative to maintain wellbore stability. In overpressured formations overburden pressure compresses the sediments, which results in fluids being expelled. This can be seen as a steady decrease in sonic slowness values, as shown in Fig. 4.25 [13]. The compaction trend in track 4 shows that there is a decrease from 135 μs/ft at X1,000 ft to 115 μs/ft at X4,400 ft. Such compaction trends are found in low permeability formations such as shale, and this is confirmed by the high gamma ray readings in track 2. Since these formations have low permeabilities

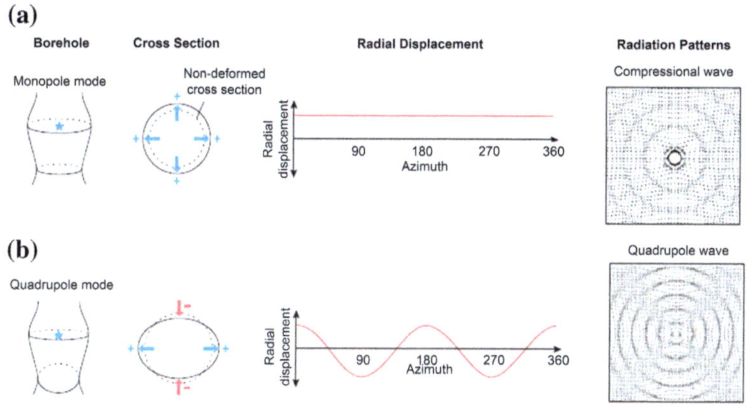

Fig. 4.24 Generation of sound waves by **a** a monopole acoustic source and **b** a quadrupole acoustic source [13]

Table 4.1 Compressional and shear wave slowness values for different materials [13]

Material	Compressional slowness time Δt_c, μs/m [μs/ft]	Shear slowness time Δt_c, μs/m [μs/ft]
Steel	187 [57]	338 [103]
Sandstone	182 [55.5]	289 [88]
Limestone	155 [47.3]	290 [88.4]
Dolomite	143 [43.5]	236 [72]
Shale	200–300 [61–91]	Varies
Freshwater	715 [218]	Not applicable
Brine	620 [189]	Not applicable

there is minimal risk of a change in the downhole pressure regime while drilling through them. However, if the fluid cannot escape the formation retains the fluid. Drilling through such formations with high porosity might lead to the hydrostatic pressure of the wellbore being insufficient to balance the pore pressure of the formation. This could result in an influx of fluids into the wellbore, which if not contained could lead to an uncontrolled flow of fluid to the surface of the well. Figure 4.25 shows how the sonic tool can potentially identify such zones by an increase in the sonic slowness from X5,000 ft in track 4 (shaded red) [13]. Therefore, drilling engineers can modify the weight of drilling fluids in real-time to counter such zones. However, the fracture thresholds of shallower formations must also be taken into account when modifying the densities of drilling fluids. It is worth to note that the resistivity measurements shown in track 3 do not give conclusive evidence pertaining to the presence of overpressured formations.

Figure 4.26 shows the LWD sonic tool measurement results in a horizontal well [13]. The sonic tool was run with resistivity (track 2), azimuthal density and neutron

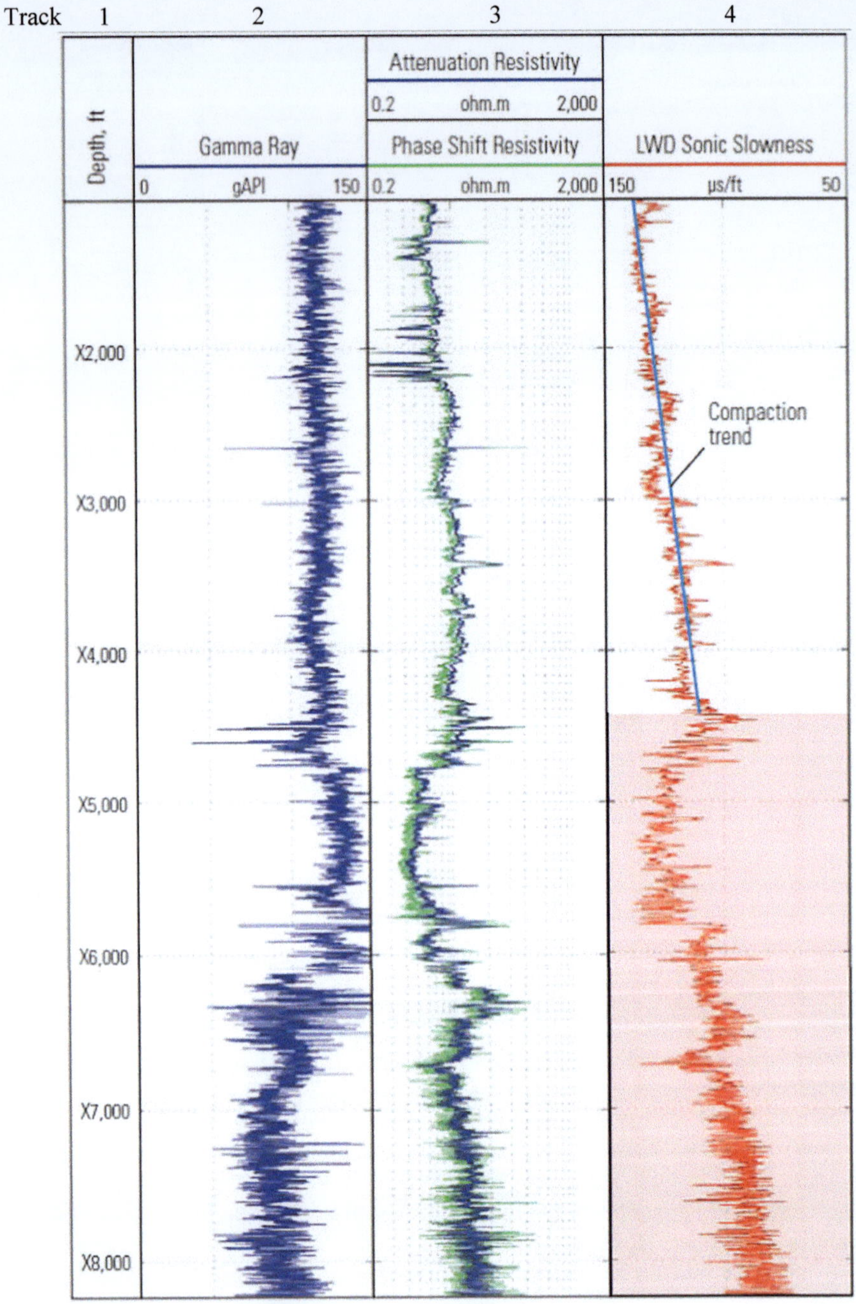

Fig. 4.25 LWD sonic data in track 4 were utilized with gamma ray measurements in track 2 and resistivity measurements in track 3 to identify compaction trends and overpressured zones. Track 1 is depth [13]

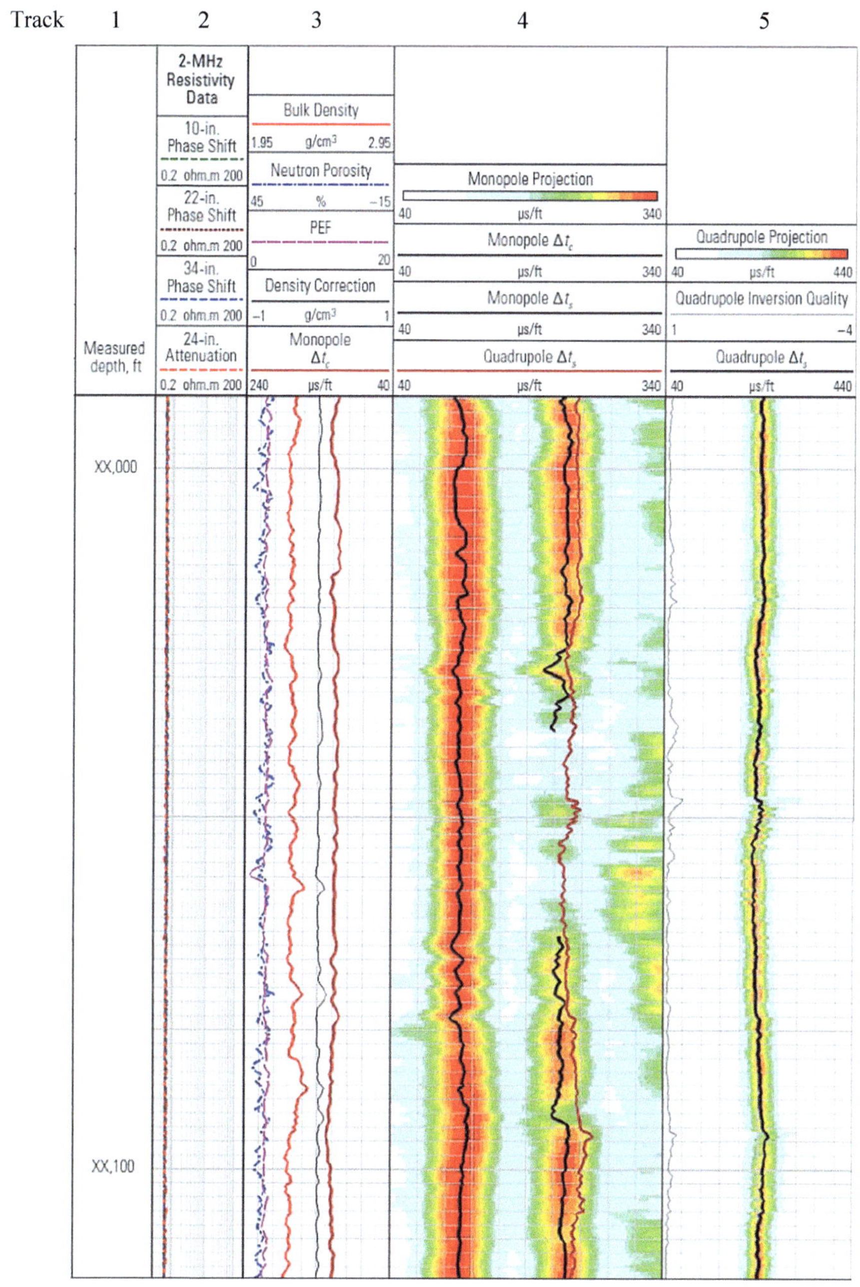

Fig. 4.26 The LWD sonic tool is able to obtain data in formations that was too slow for the monopole source as shown in track 4. Track 1 is the depth, track 2 shows the resistivity measurements, track 3 are the azimuthal density and porosity measurements and track 5 is the quadrupole data [13]

porosity (track 3). As shown in track 4, high frequency monopole and low frequency monopole data and low frequency quadrupole data were obtained by the sonic tool. There is continuous monopole compressional data throughout the logged interval (black line, track 4-left) but there is a gap between XX,038 to XX,066 ft for the monopole shear data (black line, track 4-right). This gap can be explained by the aforementioned limitation of the monopole source to only obtain shear measurements in fast formations. However, there are continuous quadrupole shear slowness measurements along the logging interval demonstrating the ability of the sonic tool to obtain data in formations that was too slow for the monopole sources. High confidence in the measurement quality is also provided by the coherence of the quadruple data in track 5. The difference between the two shear slowness measured by the two methods can be attributed to the acoustic anisotropy of the horizontal well.

References

1. B. Calleja, J. Market, J. Pitcher, C. Bilby, Multi-sensor geosteering, in *SPWLA 51st Annual Logging Symposium*, Perth, Australia (2010)
2. A. Al-Hajari, A. Soremi, S.M. Ma, A. Julaih, T.W. Thompson, G. Saghiyyah, A. Lotfy, M.A. Bayrakdar, M.S. Bittar, R.E. Chemali, Proactive geosteering in thin reservoir bound by anhydrite in Saudi Arabia, in *International Petroleum Technology Conference*, Doha, Qatar (2009)
3. M. Bittar, J. Klein, R. Beste, G. Hu, M. Wu, J. Pitcher, C. Golla, G. Althoff, M. Sitka, V. Minosyam, M. Paulk, A new azimuthal deep-reading resistivity tool for geosteering and advanced formation evaluation, in *SPE Annual Technical Conference and Exhibition*, California, USA (2007)
4. M. Bittar, J. Klein, R. Beste, G. Hu, M. Wu, J. Pitcher, C. Golla, G. Althoff, M. Sitka, V. Minosyam, M. Paulk, A new azimuthal deep-reading resistivity tool for geosteering and advanced formation evaluation. SPE Reserv. Eng. Eng. **12**, 270–279 (2009)
5. B. Anderson, T. Barber, R. Leveridge, R. Bastia, K.R. Saxena, A.K. Tyagi, J.B. Clavaud, B. Coffin, M. Das, R. Hayden, T. Klimentos, C.C. Minh, S. Williams, Triaxial induction—a new angle for an old measurement. Oilfield Rev. **20**, 64–84 (2008)
6. J. Mude, S. Arora, T. McDonald, J. Edwards, Wireline dielectric measurements make a comeback: applications in Oman for a new generation dielectric log measurement, in *SPWLA 51st Annual Logging Symposium*, Perth, Australia (2010)
7. C. Maranuk, M. Mickael, P. Zimmermann, Applications of a unique spectral azimuthal gamma ray tool unconventional reservoirs, in *SPWLA 54th Annual Logging Symposium*, Louisiana, USA (2013)
8. F. Allioli, V. Cretoiu, M.L. Mauborgne, M. Evans, R. Griffiths, F. Haranger, C. Stoller, D. Murray, N. Reichel, Formation density from a cloud, while drilling. Oilfield Rev. **25**, 4–15 (2013)
9. N. Reichel, M. Evans, F. Allioli, M.L. Mauborgne, L. Nicoletti, F. Haranger, N. Laporte, C. Stoller, V. Cretoiu, E. El Hehiawy, R. Rabrei, Neutron-gamma density (NGD): principles, field test results and log quality control of a radioisotope-free bulk density measurement, in *SPWLA 53rd Annual Logging Symposium*, Cartagena, Colombia (2012)
10. T. Kruspe, H.F. Thern, G. Kurz, M. Blanz, R. Akkurt, S. Rawaili, D. Seifert, A. Marsala, Slimhole applications of magnetic resonance while drilling, in *SPWLA 50th Annual Logging Symposium*, Texas, USA (2009)

11. C.P. Gooneratne, B. Li, T.E. Moellendick, Downhole applications of magnetic sensors. Sensors **17**, 2384 (2017)
12. S. Mohammed, J. Crowe, D. Belaud, H. Yamamoto, J.M. Degrange, V. Pistre, and H. Prabawa, Latest generation logging while drilling sonic tool: multipole acoustics measurements in horizontal wells from offshore West South Africa, in SPWLA 52nd Annual Logging Symposium, Colorado, USA (2011)
13. J. Alford, M. Blyth, E. Tollefsen, J. Crowe, J. Loreto, S. Mohammed, V. Pistre, A. Rodriguez-Herrera, Sonic logging while drilling—shear answers'. Oilfield Rev. **24**, 4–15 (2012)

Chapter 5
Downhole Communication and Power Supplies to Instruments and Communication Modules

Abstract This chapter describes the four main telemetry methods, mud pulse, electromagnetic, acoustic and wired cable, employed by downhole logging and directional instruments to communicate with the surface. The real-time transmission of data from downhole to the surface, and from the surface to downhole, enables the driller to make informed decisions and optimize the drilling process. Batteries and turbine alternators, which supply power to the logging, directional and telemetry modules, are also discussed in this chapter.

5.1 Mud Pulse Telemetry

The key enabling technology for measurement-/logging- while drilling (MWD/LWD) is real-time telemetry of data from the bottom hole assembly to the surface. Mud pulse (MP) telemetry is by far the most common method. The first commercial MP systems that appeared in the late 1970s and early 1980s were the result of research that began in the 1950s with the aim of developing a real-time communication channel between downhole instruments and the surface [1–6]. Early MP units were only able to transmit inclination, azimuth and toolface data for well navigation but gradually evolved into units capable of transmitting measurements from several different MWD/LWD instruments.

MP telemetry systems take advantage of the circulation of drilling mud to transmit data. During drilling, mud is circulated from pumps at the surface, down the drillstring, through the BHA, out the bit and back to the surface. In MP telemetry, a valve in the BHA can be momentarily closed, obstructing the mud flow and producing a back pressure pulse that can be detected at the surface. Data is represented by the timing of these pulses. Figure 5.1 shows how data from MWD/LWD is transmitted by MP telemetry to the surface. The MP unit is placed in the MWD tool and the data from the instruments in MWD/LWD tools is passed onto the microprocessor/control system of the MP unit, where the data is compressed, modulated and encoded. MP telemetry is activated by a pre-programmed mechanism such as drilling fluid flow or drilling fluid pressure increase within the

C. P. Gooneratne et al., *Instruments, Measurement Principles and Communication Technologies for Downhole Drilling Environments*, Smart Sensors, Measurement and Instrumentation 32, https://doi.org/10.1007/978-3-030-04900-3_5

drillstring assembly. The MP unit then sends pressure pulses corresponding to the instrument data to the surface, and these pressure pulses are converted to comprehensible data by pressure transducers and signal processing on the surface. The system at the surface has a matching system to demodulate, decode and decompress the data. This process is an example of 'uplink' communication. In 'downlink' communication pressure pulses are sent from the surface to the MP unit by making changes to the surface pressure, flowrate or RPM, depending on the sophistication of the particular system in use. The real-time bi-directional data is used for timely decisions when drilling, for example to control the steering and direction of the drilling assembly to maximize drilling in highly permeable hydrocarbon formations. Over the last two decades MP systems have increased their performance to accommodate advancements made in logging instruments, such as high resolution wellbore images, directional instruments, such as RSSs, and drilling dynamics instruments, such as vibration sensors.

MP telemetry is an electromechanical system that creates pressure waves within the drilling fluid through either positive, negative or continuous pulses, as shown in Fig. 5.1b. Early MP units used positive and negative pulses, where the typical waveform is a pulse and the process in known as pulse modulation. In positive pulse telemetry a valve blocks and unblocks the drilling fluid flow inside the drillstring for short times so that the pressure increases and then returns to its original state, as shown in Fig. 5.1b-(i). Initially reference pulses are sent to the surface by operating the valve several times followed by data pulses corresponding to instrument measurements. The reference pulses are identified by the surface transducers, and the data pulses that follow the reference pulses within a particular time-frame are converted from binary code into a decimal result. In negative pulse telemetry a valve is used to divert drilling fluid from inside the drilling tool to the wellbore thus, reducing the pressure inside the drilling tool, as shown in Fig. 5.1b-(ii). This pressure drop is recognized at the surface transducers and data processing similar to positive mud pulse telemetry is performed at the surface. In continuous pulse telemetry a stator/rotor system is used to produce continuous positive pressure pulses, as shown in Fig. 5.1b-(iii). Continuous pulse telemetry employs passband modulation where the signal is encoded in the frequency or the relative phase of the signal. The frequency of the generated wave is proportional to the rotation speed of the rotor.

Positive, negative and continuous MP telemetry methods deliver strong, discrete pulses or low, amplitude continuous waves with a maximum data rate around 3 bps. However, the continuous addition of new MWD/LWD instruments significantly increased the data density that needed to be transmitted across the low bandwidth telemetry channel [7–12]. Some methods to resolve this challenge included faster pulsers, more efficient encoding, data compression techniques as well as reducing the rate of penetration but these methods were found to be unfeasible as wells became more complex, data density requirements continued to increase and rig rates kept rising.

A new MP unit to meet this challenge was first used in the late 2000s. The new method includes an oscillating pulser, which has a shear valve as a stator and an

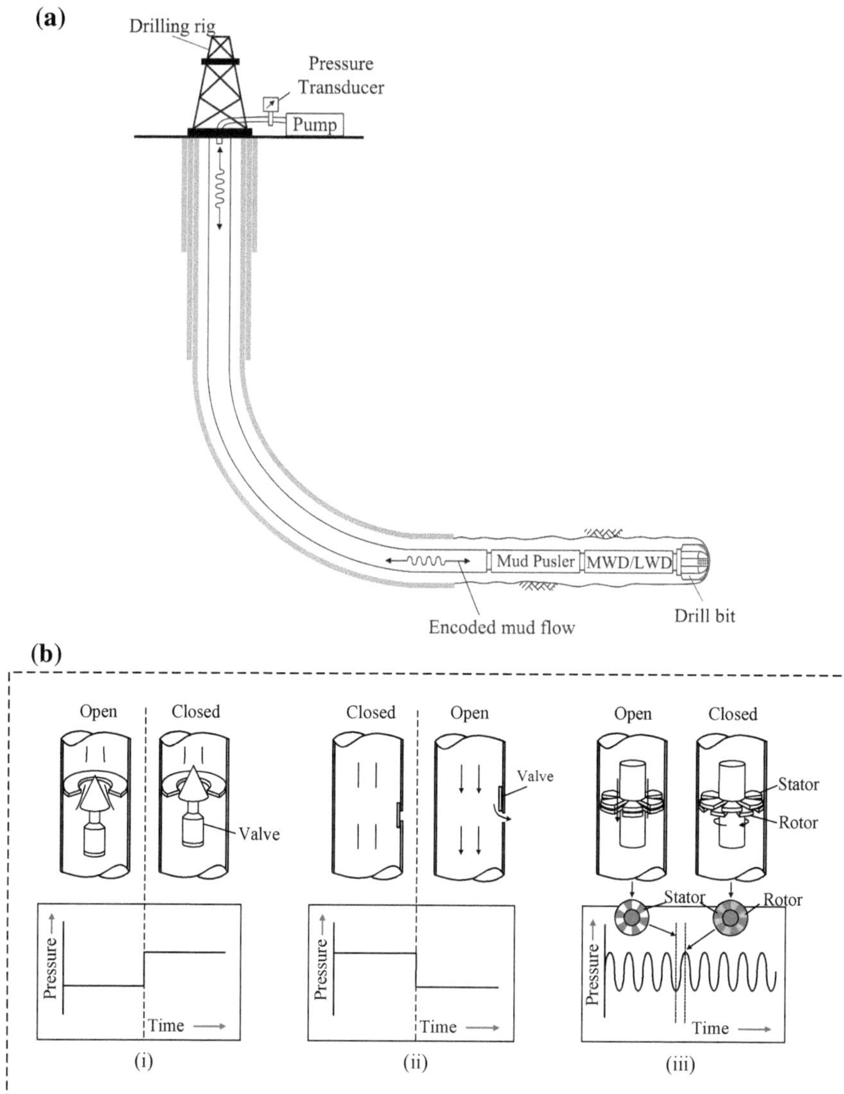

Fig. 5.1 **a** A mud pulse telemetry system, where data from the MWD/LWD to the surface and from the surface to the MWD/LWD is transmitted using pressure pulses created in the drilling fluid. **b** The three main methods of creating pressure pulses. (i) Positive pulses by blocking and unblocking the drilling fluid flow using a valve, (ii) negative pulses by releasing pressure inside the drillstring through a valve that opens to the wellbore, and (iii) continuous pulses by a stator/rotor system

oscillating rotor to produce continuous positive pressure pulses [13–16]. The main difference between the previous continuous MP and this method is that the rotor is not fixed and can oscillate in the oscillating pulser. Therefore, signaling flexibility is provided by being able to generate both discrete wave signals like positive and negative MP and continuous wave signals like continuous MP. Other advantages include low power consumption at high speeds of up to 20 bits/s, instantaneous change of phase or frequency for efficient telemetry modulation, easy adaptation to drilling fluid flow rate and changes in fluid density from the surface through downlink signals, reduction of higher harmonics of signals, full control of the shape of the pressure pulse and general adaptability of the oscillating pulser to the environment [17–23].

Even with current progress in MP telemetry there are still several noise sources that affect signal quality since the drilling environment is dynamic and unique. Increasing depth decreases the amplitude of the signal, high compression drilling fluids such as oil based fluids attenuate the signal, drilling noises such as the noise created by torque of the drillstring assembly when the drill bit penetrates through formations, drillstring reflections and noise at the surface from drilling fluid pumps and drilling motors affect the signal to noise ratio of the signal. Depending on the well situation and drilling complexity several techniques such as increasing the flow rate, decreasing the drilling fluid viscosity, high-pass/low-pass/band-pass/matched filtering techniques, pump noise cancellation algorithms, drillstring design practices and optimization models can be used to mitigate noise and decrease signal attenuation.

Since MP telemetry uses mud as a medium it is highly influenced by the type of drilling fluid and cannot be used when drilling with foam or air due to their high compressibility. MP telemetry propagates pressure waves at speeds in the range of 1000–1500 m/s and while its maximum data bandwidth is 20 bits/s, in most cases it is still limited to 3–6 bits/s. While this is a significant improvement from the early 0.4 bps MP units, the rate of development of current units has been outpaced by the rapid progress in logging, directional and drilling dynamics instruments, which produce increasingly higher resolution data. Therefore, higher bandwidth data communication channels are required to transmit higher volumes of data to the surface. This will provide the driller at the surface more accurate drilling dynamics data and better quality maps of the rock formations and hydrocarbon reservoirs, and aid in the precise navigation of wells.

5.2 Electromagnetic Telemetry

Electromagnetic (EM) telemetry uses low frequency electromagnetic waves in the region of 2–20 Hz for bi-directional communication [24–27]. Figure 5.2 shows a general schematic of an EM system with the drillstring as a dipole electrode transmitter and two antennas located at the surface as receivers. A voltage is generated between the drillstring below the insulator and the drillstring above the

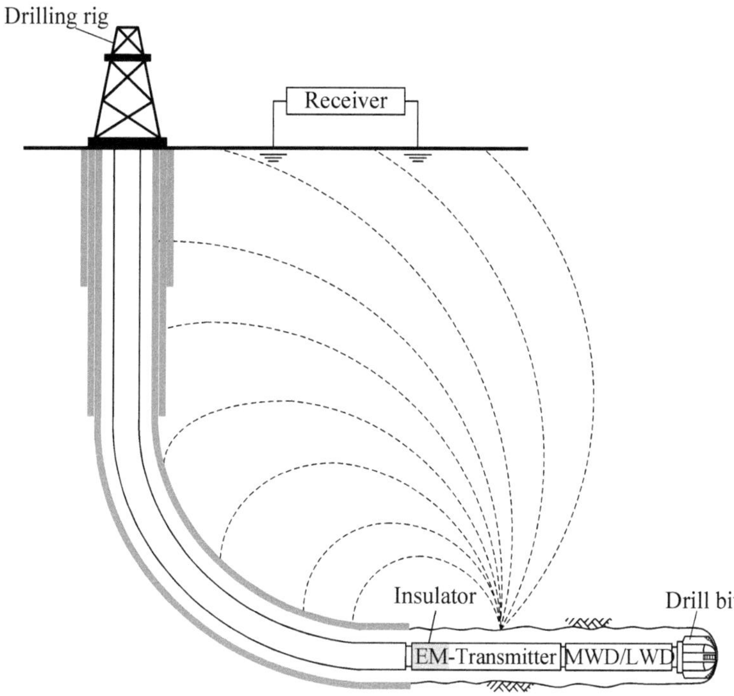

Fig. 5.2 Data communication from the MWD/LWD to the surface and from the surface to the MWD/LWD by electromagnetic telemetry. For data transmission from the MWD/LWD unit to the surface, a voltage applied above and below the insulator makes the drillstring a dipole electrode transmitter that transmits electromagnetic waves and the two antennas on the surface a dipole antenna that receives the electromagnetic wave signals. The opposite is true when data needs to be sent from the surface to the MWD/LWD unit

insulator. One antenna at the surface is connected to the top of the drillstring assembly and the other is driven into the surface some distance away from the drillstring assembly. These two antennas form a dipole antenna that receives the electromagnetic wave signals transmitted through rock formations from downhole. EM telemetry can reach data speeds up to 6 bits/s but is dependent on formation and drilling fluid resistivity and frequency of transmission. Compared to MP telemetry, EM telemetry has no moving parts downhole, and is independent of drilling fluid flow and lost circulation material. EM telemetry is mostly utilized in underbalanced drilling operations, where drilling is performed with air or foam instead of drilling fluids to allow hydrocarbons to flow to the surface while drilling, or when drilling onshore wells since it is challenging to place receivers on the seabed in offshore operations and water being an electrical conductor attenuates the electromagnetic propagation [27–36]. The main limitation of EM telemetry is the severe signal attenuation due to skin depth effects [37, 38]. While lowering the frequency can improve propagation through formations it also results in lower data rates. Moreover, the conductivity of the drilling fluid also influences signal transmission and signal attenuation increases exponentially with increasing well depth. Therefore, most EM systems are used in shallow wells with depths up to around 4000 m even though depths of up to 5000 m have been reached in some wells. An optimal environment for EM telemetry is one where the drilling fluid and formation resistivity is low enough to inject a signal into the formation but where the formations above leading to the surface has a high enough resistivity to minimize signal attenuation.

Several approaches to improve EM telemetry have been implemented over the years. Repeaters have been utilized to improve the data rates and extend the depth range. Measuring the potential difference between bottom of the cable and the wellhead of an adjacent well, using novel casing and antenna systems, and insulated wire externally connected to a standard casing string, where a wellbore receiver is connected to the casing and a surface transceiver, have also been employed to improve data rates and extend EM telemetry to long lateral wells [37–42]. Moreover, it also has applications in pad drilling, where one well in a drilling pad acts as an anchor and serve as a receiving antenna for other wells drilled on the pad.

5.3 Acoustic Telemetry

Patents on acoustic telemetry date back to the 1940s and research into the propagation of acoustic energy for telemetry purposes in a drilling environment started in the 1970s–80s [43–50]. Acoustic telemetry uses the steel walls of a drillstring assembly as the main communication channel to transmit sound waves to the surface, at frequencies ranging from 400 Hz to 2 kHz, where a receiver receives and decodes the data. Figure 5.3 shows an acoustic telemetry system with a downhole piezoelectric transmitter with electronics and a battery, repeaters that are normally spaced 500–2000 m apart, and a receiver at the surface, which typically consists of two accelerometers, a processor and associated electronics for signal filtering and amplification, a wireless transceiver for bi-directional communication with the decoding and display unit, and a battery pack. While technically data rates can be very high, repeaters are generally required since there is significant attenuation when the sound waves travel up the drillstring assembly, especially at drill pipe connections, where the signal is partially reflected and partially transmitted, and at any location where the drillstring assembly is in contact with the wellbore [51–54]. A repeater receives a signal from the repeater located below it, which it then decodes and re-transmits to the repeater above it. This process is repeated until the signal is transmitted all the way to the receiver at the surface. Repeaters not only boost the signal but can also take measurements of the local pressure and temperature so real-time pressure/temperature profiles are available to the driller at the

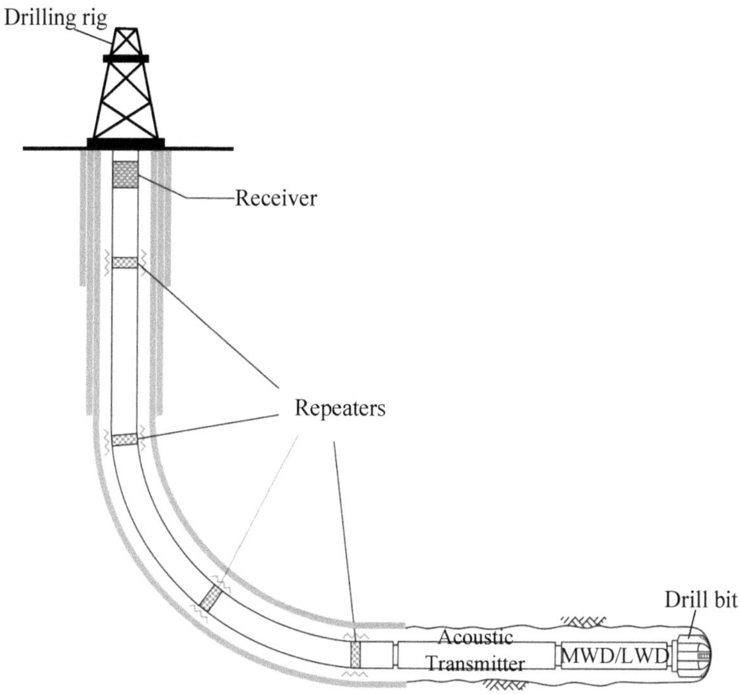

Fig. 5.3 Acoustic telemetry, where data is transmitted up and down a drillstring by sound waves and repeaters act as signal processors and data amplifiers

surface, which can be utilized to optimize drilling. Moreover, mechanical parameters such as weight and torque can be recorded for understanding the effects these parameters have on downhole problems. Field trials have shown that downhole data can be obtained and replicated during cementing, casing and completion related operations in shallow and underbalanced wells [55–57].

Even though repeaters are used to increase the depth range and the signal strength the failure of a repeater or repeaters along the drillstring could sever the communication link. Also, acoustic signal losses in directional wells are higher than in vertical wells because there are more points of contact between the drillstring assembly and the wellbore, which is further amplified by the rock formation cuttings on the low side of the well. Finally, and most importantly, extending the success of obtaining pressure data along a wellbore in a non-rotating environment to successfully transmitting data from a range of MWD/LWD tools in a high noise drilling environment is a formidable task. Therefore, further development is needed to use acoustic telemetry for transmitting data from different MWD/LWD tools in both shallow and deep wells.

5.4 Wired Cable Data Telemetry

Wired telemetry uses special drill pipe with an embedded cable that conveys the telemetry signal along pipe joints and inductive coils that couple the signal across the connections between pipe joints [58–60]. Such a direct connection allows bi-directional data transmission rates up to 57,600 bps, which is five orders of magnitude higher than the data rates achieved by other wireless methods. This high bandwidth allows wired telemetry to obtain large amounts of high quality data from MWD/LWD tools, and for example, rapidly send information based on this high resolution data to an RSS for precise well navigation.

Figure 5.4 shows a wired telemetry system with four main components, an interface sub, repeaters, wired cable pipe, and a top drive swivel. The interface sub provides a physical and electrical crossover that connects the MWD/LWD tools and the RSSs to the wired system. The wired cable pipe is a specially manufactured drill pipe to house the high strength, high speed coaxial cables and the low-loss inductive coils as shown in the inset in Fig. 5.4. The drill pipes are modified by having a channel etched on the side of the pipe wall to embed the cable under tension. The cable is terminated as a loop at either side of the drill pipe and is anchored in a groove, which is a continuation of the side channel on the wall of the drill pipe. Therefore, when two drill pipes are connected together the two loops will be brought in close proximity and opposite to each other, as shown in the inset in Fig. 5.4. This configuration allows a current flowing in a coaxial cable in a drill pipe to transfer the current to a coaxial cable in a drill pipe above or below it by electromagnetic induction, where the electromagnetic field associated with the current in one coil induces a current in the nearby coil [61]. This process is repeated to transfer data from bottomhole to the surface or in the opposite way. The electromagnetic method of data transfer is much simpler than running a single cable through all the drill pipes with slots at each drill pipe to connect the cables, since this requires precise orientation of the drill pipes. Moreover, connected cables can easily be disconnected during the rotation of the drillstring assembly. The repeaters are utilized to boost the current signal, which attenuates as it travels up the string, and also as an along the drillstring measurement system, where instruments to measure parameters such as temperature, pressure and drilling dynamics data can be incorporated into the repeaters to obtain real-time measurement profiles for the entire well; this can be done in any EM or acoustic telemetry system too. Therefore, wired telemetry can be useful in applications such as managed pressure drilling, underbalanced drilling and drilling automation [62–68]. The data transmitted from bottomhole is extracted by a swivel located at the top drive and provides the interface between the rotating drillstring assembly and the stationary data extraction module.

Even though wired telemetry offers significantly higher data rates than wireless telemetry methods, the extensive modifications required to be made to a standard drillstring assembly and a drilling rig make it a very expensive operation. Since wired telemetry is a system with many parts, handling wired telemetry equipment

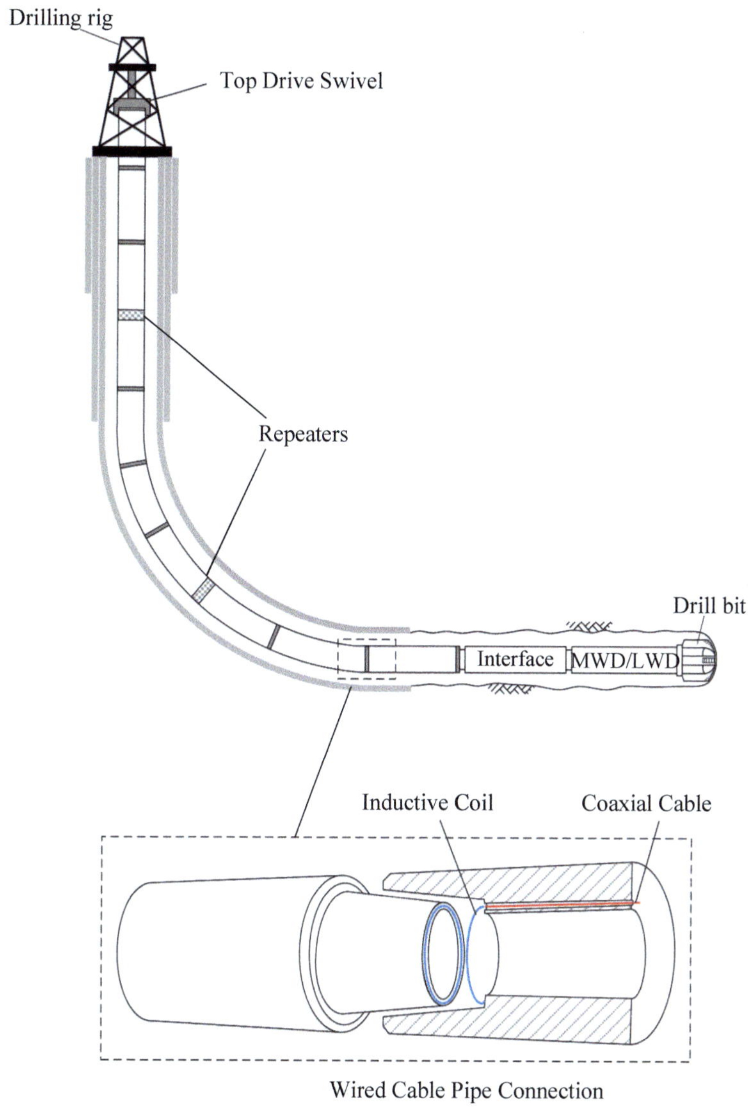

Fig. 5.4 Wired telemetry, where a wired cable extending from the surface to the bottom of the well provides real-time, bi-directional data transmission. Data is transferred from one drill pipe to another by electromagnetic induction and repeaters are used as signal processors and amplifiers

requires qualified personnel and training will also have to be provided to the crew handling the connections of the drill pipes on the specific rig so as not to damage the embedded wired cables or dislodge them from the trenches and grooves. The repeaters are powered by batteries with finite lifetimes before they have to be replaced or recharged so a repeater failing during an operation due to a battery being fully discharged or some other issue will affect the data rate and quality.

While co-axial cables are well insulated and armored, they are constantly exposed to a severe environment of drilling fluids, high downhole temperatures and pressures and possible improper handling by the rig crew, and are therefore susceptible to failure. In the event of a repeater or a cable failure the drillstring assembly will have to be pulled out of the well to perform maintenance on or replace the repeater or cable. The cost of the wired drill pipe and the risk of a failure will increase with the depth of the well. Due to these issues wired telemetry is currently not widely used when drilling wells.

5.5 Power Supplies for Downhole Instruments

5.5.1 Batteries

LWD/MWD tools do not have an electrical cable providing power to them from the surface as wireline (WL) tools do. They need a downhole power source. Drilling imposes several additional restrictions on a downhole power source, such as high temperature, high pressure, shock, vibration and limited space in a drillstring [69]. Lithium (Li) batteries most commonly supply power to the microprocessors/microcontrollers, signal processing and electronics rather than the measurement instruments and data communication units, because batteries have limited instantaneous energy output and are unsuitable for applications that require a high current drain. It is preferable to use Li batteries with the highest temperature tolerance so Li primary batteries, which have lithium metal or lithium compounds, rated at 175 °C are used instead of secondary batteries, known as Li-Ion, which have a temperature tolerance of 125 °C [70]. Primary batteries unlike secondary batteries are not rechargeable so are one-time use and disposable. Li batteries are defined by their shelf and operating lives, where the former refers to their date of expiry and the latter depends on the size of the battery, power management and the environments they are used in.

Lithium thionyl chloride batteries with a voltage of 30 V are the most used downhole batteries because they have a high energy density, are low weight and only take up a small space in the drillstring [3, 69]. However, they melt at 180 °C, so if the batteries are run beyond this limit the inner and outer temperature of the battery runs the risk of melting the lithium metal, which can cause a violent, accelerated reaction and explode with a force large enough to create a hole through the pressure housing and damage the tool [70]. Magnesium can be alloyed with Li to increase the operating temperature but this results in a reduction in current capacity. All types of batteries discharge over time and high temperatures downhole accelerate this process [71–74]. Batteries are also expensive, require maintenance or replacement and have the added cost of safe disposal due to the chemicals they contain.

5.5.2 Turbine/Alternator

The most power consuming parts of MWD/LWD tools are the acquisition of measurements by different instruments and the transmission of these measurements as data to the surface. A turbine alternator (TA) uses drilling fluid flow to generate electricity and power downhole instruments and communication systems [75, 76]. The generated power is proportional to the flow rate of the drilling fluid and therefore, unlike batteries there are no hour limits to the available power as long as there is fluid circulation in the well. In the basic operation of a turbine alternator system the kinetic energy of a drilling fluid flow rotates a turbine and this rotational force is transferred by a shaft to an alternator. The alternator converts this rotational force to a three phase ac current with a frequency that is dependent on the rate of rotation. The generated ac current is converted to a dc current by a rectification circuit to recharge a downhole battery. Typical features include the ability to generate 200–300 W and operate at 10,000–15,000 psi and a temperature of 150 °C continuously for 300–400 h, where the power required depends on the number of measurement instruments in the MWD/LWD tools and the data transmission rates [77]. TA can operate for around 300–400 h at a temperature of 150 °C.

There is a minimum drilling fluid flow threshold for the TA to generate power and there will be no power generated in the absence of drilling fluid flow in the well [78]. Li batteries are independent of drilling fluid flow so are able to act as a temporary power source if needed to provide limited power to measurement instruments when there is no circulation in the well. The turbine may be damaged if the flow exceeds a maximum threshold and turbine is also susceptible to damage by large particles or debris in the drilling fluid and lost circulation material. Generally a screen is installed above the TA system, at the top of the drillstring assembly, to filter such particles or debris reaching the turbine. The screen is placed in a position where it can be periodically cleaned or removed if it prevents a tool access into the drillstring assembly.

References

1. B.J. Patton, W. Gravley, J.K. Godbey, J.H. Sexton, D.E. Hawk, V.R. Slover, J.W. Harrell, J. Petrol. Technol. **29**, 1215 (1977)
2. E.B. Denison, J. Petrol. Technol. **31**, 155 (1979)
3. M. Gearhart, K.A. Ziemer, O.M. Knight, J. Petrol. Technol. **33**, 2301 (1981)
4. A. Roberts, R. Newton, F. Stone, in *SPE Annual Technical Conference and Exhibition* (1982)
5. W. Gravley, J. Petrol. Technol. **35**, 1439 (1983)
6. J.B. Cheatham Jr., in *SPE Production Technology Symposium* (1983)
7. C.G. Blount, M.B. Mooney, B.E. Smith, D. Quinn, E.B. Larson, in *SPE Annual Technical Conference and Exhibition* (1997)
8. R. Hutin, R.W. Tennent, S.V. Kashikar, in *SPE/IADC Drilling Conference* (2001)
9. Q. Li, T. Bornemann, J. Rasmus, H. Wang, K. Hodenfield, J. Lovell, in *SPWLA 42nd Annual Logging Symposium* (2001)

10. M.J. Sullivan, J.R. Scheevel, B. Katz, in *SPWLA 43rd Annual Logging Symposium* (2002)
11. J. Horkowitz, S. Crary, K. Ganesan, R. Heidler, B. Luong, J. Morley, M. Petricola, C. Prusiecki, M. Poitzsch, J.R. Scheibal, M. Hashem, in *SPWLA 43rd Annual Logging Symposium* (2002)
12. O.N. Akimov, A. Baule, J. Bethge, T. Roessel, R. Tilsley-Baker, in *SPE Europec/EAGE Annual Conference and Exhibition* (2006)
13. C. Klotz, I. Wassermann, D. Hahn, in *SPE Indian Oil and Gas Technical Conference and Exhibition* (2008)
14. C. Klotz, P.R. Bond, I. Wassermann, S. Priegnitz, in *SPE/IADC Drilling Conference* (2008)
15. C. Klotz, A. Kaniappan, A.K. Thorsen, E. Nathan, M. Jahangir, L. Lie, in *SPE/IADC Asia Pacific Drilling Technology Conference and Exhibition* (2008)
16. I. Wasserman, D. Hahn, D.H. Nguyen, J. Macpherson, Oil Gas J. **106**, 39 (2008)
17. P. Cooper, L.S.B. Santos, in *SPE Russian Petroleum Technology Conference* (2015)
18. W. Emmerich, O. Akimov, I.B. Brahim, A. Greten, in *SPE/IADC Drilling Conference and Exhibition* (2015)
19. W. Emmerich, O. Akimov, I.B. Brahim, A. Greten, in *SPE/IADC Drilling Conference and Exhibition* (2016)
20. W. Emmerich, I.B. Brahim, O. Akimov, A. Greten, in *Offshore Technology Conference Asia* (2016)
21. W. Emmerich, A. Greten, I.B. Brahim, O. Akimov, in *Offshore Technology Conference* (2016)
22. A. Greten, I.B. Brahim, W. Emmerich, O. Akimov, in *SPE/IADC Drilling Conference and Exhibition* (2017)
23. Z. Yan, Y. Geng, C. Wei, T. Wang, T. Gao, J. Shao, X. Hu, M. Yuan, Flow Meas. Instrum. **59**, 28 (2018)
24. D.A. Hill, J.R. Wait, Electron. Lett. **14**, 532 (1978)
25. J. Bhagwan, F.N. Trofimenkoff, Electric drill stem telemetry. IEEE Trans. Geosci. Remote Sens. **GE-20**, 193 (1982)
26. P. DeGauque, R. Grudzinski, SPE Drill. Eng. **2**, 127 (1987)
27. L. Soulier, M. Lemaitre, in *SPE/IADC Drilling Conference* (1993)
28. W.H. Harrison, R.L. Mazza, L.A. Rubin, A.B. Yost, in *SPE/IADC Drilling Conference* (1990)
29. F.N. Trofimenkoff, M. Segal, A. Klassen, J.W. Haslett, IEEE Trans. Geosci. Remote Sens. **38**, 2539 (2000)
30. D. Weisbeck, G. Blackwell, D. Park, C. Cheatham, in *SPE/IADC Drilling Conference* (2002)
31. B. Peter, D. Weisbeck, R. Graham, in *SPE/IADC Underbalanced Technology Conference and Exhibition* (2003)
32. P.K. Vong, D. Rodger, A. Marshall, IEEE Trans. Magn. **41**, 2008 (2005)
33. S. Wilhide, J. Smith, D. Doebereiner, B. Raymond, D.H. Weisbeck, B. Ziemke, in *SPE Annual Technical Conference and Exhibition* (2010)
34. G. Kolaric, J. Lee, S. Yong, K. Yu, G. Fu, S. Tseytlin, in *Offshore Mediterranean Conference and Exhibition* (2011)
35. L. Serniotti, M. Troiano, D. Di Tommaso, in *Offshore Mediterranean Conference and Exhibition* (2011)
36. C. Carpenter, J. Petrol. Technol. **67**, 87 (2015)
37. M.Y. Xia, Z.Y. Chen, IEEE Trans. Geosci. Remote Sens. **31**, 1222 (1993)
38. J. Schnitger, J.D. Macpherson, in *SPE/IADC Drilling Conference and Exhibition* (2009)
39. J. Chen, S. Li, C. MacMillan, G. Cortes, D. Wood, in *SPE Annual Technical Conference and Exhibition* (2015)
40. Q. Dong, J. Chen, in *SEG International Exposition and Annual Meeting of the Society of Exploration Geophysicists* (2016)
41. G. Jannin, J. Chen, L.E. DePavia, L. Sun, M. Schwartz, in *SEG International Exposition and Annual Meeting of the Society of Exploration Geophysicists* (2017)
42. S. Zeng, Q. Dong, J. Chen, in *Unconventional Resources Technology Conference* (2017)

43. T.G. Barnes, B.R. Kirkwood, J. Acoust. Soc. Am. **51**, 1606 (1972)
44. A.H. Drouin, in *Offshore Technology Conference* (1974)
45. D.R. Witmer, R.E. Pearson, in *Offshore Technology Conference* (1977)
46. W.D. Squire, H.J. Whitehouse, in *SPE Annual Technical Conference and Exhibition* (1979)
47. C.H. Cheng, M.N. Toksöz, Geophysics **46**, 1042 (1981)
48. R.J. Trett, in *Offshore Technology Conference* (1988)
49. D.S. Drumheller, J. Acoust. Soc. Am. **85**, 1048 (1989)
50. T. Tochikawa, T. Sakai, R. Taniguchi, T. Shimada, in *SPE Annual Technical Conference and Exhibition* (1996)
51. L. Gao, W.R. Gardner, C. Robbins, D.H. Johnson, M. Memarzadeh, SPE Reserv. Eval. Eng. **11**, 141 (2008)
52. P.L. Camwell, J.G. McRory, J.M. Neff, Drill. Contractor **65**, 30 (2009)
53. M.E. Reeves, P.L. Camwell, J. McRory, in *SPE Offshore Europe* (2011)
54. A.K. Farraj, S.L. Miller, K.A. Qaraqe, in *SPE Annual Technical Conference and Exhibition* (2013)
55. W.R. Gardner, R.E. Hyden, E.J. Linyaev, L. Gao, C. Robbins, J. Moore, in *SPE/IADC Drilling Conference* (2006)
56. J.M. Neff, P.L. Camwell, in *SPE/IADC Drilling Conference* (2007)
57. A. Hawthorn, S. Aguilar, in *SPE Annual Technical Conference and Exhibition* (2017)
58. M. Reeves, J.D. MacPherson, R. Zaeper, D.R. Bert, J. Shursen, W.K. Armagost, D.S. Pixton, M. Hernandez, in *SPE/IADC Drilling Conference* (2006)
59. M. Hernandez, D.W. MacNeill, M. Reeves, A.D. Kirkwood, J.P. Ruszka, R. Zaeper, S.R. Lemke, in *SPE Indian Oil and Gas Technical Conference and Exhibition* (2008)
60. R.B. Russell, M. Hernandez, D.W. MacNeill, M.E. Reeves, P. Hamel, in *International Petroleum Technology Conference* (2008)
61. A.D. Craig, T.A. Jackson, D. Ramnarace, R. Schultze, M. Briscoe, in *SPE/IADC Drilling Conference* (2013)
62. M.J. Jellison, D.R. Hall, D.C. Howard, H.T. Hall Jr., R.C. Long, R.B. Chandler, D.S. Pixton, in *SPE/IADC Drilling Conference* (2003)
63. J.E. Gravdal, in *International Petroleum Technology Conference* (2009)
64. A.E. Gravdal, R.J. Lorentzen, R.W. Time, in *SPE Asia Pacific Oil and Gas Conference and Exhibition* (2010)
65. S.T. Edwards, C.J. Coley, N.A. Whitley, R.G. Keck, V. Ramnath, T. Foster, K. Coghill, M. Honey, in *SPE/IADC Drilling Conference* (2013)
66. D.S. Pixton, R.A. Shishavan, H.D. Perez, J.D. Hedengren, A. Craig, in *SPE/IADC Managed Pressure Drilling & Underbalanced Operations Conference & Exhibition* (2014)
67. H.B. Siahaan, K.S. Bjorkevoll, J.E. Gravdal, in *SPE Intelligent Energy Conference & Exhibition* (2014)
68. S. Grymalyuk, M. Schimanski, L. Lilledal, in *Abu Dhabi International Petroleum Exhibition & Conference* (2016)
69. D. Hensley, M. Milewits, W. Zhang, Oil Field Rev. **10**, 42 (1998)
70. R. Beckwith, J. Petrol. Technol. **65**, 42 (2013)
71. B.A. Stroehlein, in *Abu Dhabi International Petroleum Exhibition and Conference* (2014)
72. M. Milewits, in *The Seventeenth Annual Battery Conference on Applications and Advances* (2002), pp. 199–204
73. M.R. Palacín, A. De Guibert, Science **351**, 1253292 (2016)
74. K. Zhao, Y. Cui, Extreme Mech. Lett. **9**, 347 (2016)
75. H. Guo, Z. Lv, Z. Wu, B. Wei, IET Electr. Power Appl. **7**, 214 (2013)
76. H. Lu, B. Li, X. Zhang, IJEART **2**, 16 (2016)
77. B. Plamp, *Development of a High Pressure/High Temperature Down-Hole Turbine Generator* (Dexter Magnetic Technologies, Incorporated, 2007). DE-FC26-05NT42655, phase I final report. Available online at https://www.osti.gov/servlets/purl/903397
78. Z. Wang, Z. Jian, Y. Guo, W. Zhu, H. Wang, in *The 24th International Ocean and Polar Engineering Conference* (2014)

Chapter 6
Fabrication and Packaging of Downhole Instruments

Abstract This chapter provides an overview of the fabrication and packaging methods for downhole environments. Operation in harsh and critical environments requires instruments to meet stringent safety standards. Therefore, all layers of a downhole instrument, such as electronic circuits, components, integrated circuit modules along with encapsulation and packaging, are carefully designed, planned and tested for safe and optimal operation in a well.

6.1 Harsh Environment Challenges

Increasing energy demand and declining resources have resulted in hydrocarbon wells being drilled in increasingly harsh environments with higher pressures, temperatures, and vibrations, pushing the limits of technology [1–4]. Failed tools inside wells thousands of meters underground can take up to a day to recover, which can result in significant financial losses. All measurement-/logging- while drilling (MWD/LWD) tools and downhole communication units have sensors and electronics (S&E), which fatigue and eventually fail in harsh downhole conditions. Therefore, fabrication and packaging is not only done to extend the reliability and operation of tools but also to ensure it meets the high standards of safety required in the oil/gas industry.

Fabrication and packaging for downhole environments is challenging because the majority of electronic components are made for commercial use, while the conditions experienced in drilling are similar to niche applications such as defence, space and aeronautics [2, 4–6]. However, the demand for, high pressure, high temperature and high severity (HP/HT/HS) sensors and electronics (S&E) in space, defence and aeronautics dwarfs the need for such components in drilling. Moreover, the research into HP/HT/HS tools is also driven by market demand so research and development budgets for space, defence and aeronautics far exceed oil/gas company budgets for developing similar tools for drilling. While sharing knowledge with these applications would be ideal for oil/gas companies the nature of the jobs performed in space, defence, and to a lesser extent, aeronautics, prevent such collaborations, stifling the development of HP/HT/HS tools for drilling. Therefore,

C. P. Gooneratne et al., *Instruments, Measurement Principles and Communication Technologies for Downhole Drilling Environments*, Smart Sensors, Measurement and Instrumentation 32, https://doi.org/10.1007/978-3-030-04900-3_6

the economics of evaluating S&E for reliability in HP/HT/HS drilling environments is very high, the tools are expensive and the daily cost for such tools on a drilling rig are considerably higher than conventional tools.

S&E are isolated in a special housing or vessel to protect them from the high pressures in a well, which is typically 10,000–15,000 psi. The housing or the vessel also prevent the S&E from being directly exposed to well fluids and the uneven wellbore [7–10]. Due to the restrictions placed by the exceedingly narrower wells drilled with increasing depth, the housing/vessel is very narrow and covers a chassis that holds all the S&E. The housing/vessel is typically made of a material such as Inconel, Titanium, Nitronic 50 or Hastealloy and have O-ring pressure seals to resist pressures. Therefore, high quality mechanical design of the housing/vessel enables the S&E to be kept at atmospheric pressure while drilling in high pressure zones. Shock and vibration, depending on the magnitude, can cause acute damage and immediate failure of S&E. Therefore, mounts and valves are installed in the housing/vessel to mitigate and isolate shock/vibration. Sacrificial bumps under the housing/vessel or elastometric material on the sides of the tool can also be utilized to mitigate shock/vibration.

Temperature is arguably the single most universally influential variable that contributes to tool failures downhole, and it also sets the stage for other variables such as shock and vibration to accelerate damage to S&E [7–10]. Temperature forces S&E to act beyond their specified design limits, is a catalyst for chemical reactions in batteries, and stimulates corrosion in S&E, pressure housing/vessels and drill pipes. Temperature cannot be easily isolated like pressure but can be protected for short periods of time with dewar or vacuum flasks. This is an effective measure in wireline (WL) tools since operations typically last 5–10 h but not in MWD/LWD tools that are downhole for much longer times. The S&E are therefore designed taking this into consideration.

6.2 Fabrication and Packaging Methods

6.2.1 Electronic Circuit Design and Component Testing

De-rating S&E is often the first approach taken when designing high temperature S&E [7]. De-rating is performed to address electrical and thermal stresses an electronic component is subject to. In the downhole environment this involves reducing the electrical stress by deliberately operating the device below its maximum capability to indirectly reduce thermal stress and prolong its life even though the operating time at elevated temperatures will be reduced compared with standard temperatures. Finding S&E from a large number of commercial S&E that might work beyond their specifications is a time-consuming process due to the extensive reliability tests that needs to be performed at high temperatures. Another issue is the different temperature ratings for different components of the S&E unit, which are expected to function as an interlinked unit. This adds complexity when evaluating the reliability of S&E to work as a unit because the overall temperature rating of an electrical circuit is less than the rating of the weakest component in the circuit. Therefore, special circuit designs have to be made to incorporate all the different S&E components into one system and then characterized to test its reliability. The objective when designing these circuit boards is to always include the minimum number of S&E to minimize the number of failure points.

Removing all organic chemicals, such as plastic packaging, from the S&E and hardening the electronics by increasing the activation energy required to start a chemical reaction is also an effective way to manage temperatures [6–8, 10, 11]. Excess temperatures for example would cause the bromin in plastic packaging to release and then start corroding, leading to the failure of S&E [12]. Also, space is at a premium inside the narrow drill pipe so it is not feasible to use leaded, through-hole packages to counter vibration and thermal expansion of circuit boards, so a shift was made in the industry to use multi-chip modules (MCM), which can house several components, to not only optimize the space available in drill pipes to get higher density and reliability but also to move away from organic chemicals that cause corrosion [7, 9–12]. Cooling, other than flasking explained before, include cold plates and refrigeration [8]. In cold-plate techniques, which are commonly used in personal computers for heat management, a current is passed through metallic plates to reduce their temperature and attract heat away from S&E. Flasking involves inserting the S&E into a chamber in a vacuum state or filled with gas thereby reducing the heat transfer from the environment to the electronics. Refrigerators are also used for cooling S&E by pumping a liquid, which evaporates and becomes colder as it is allowed to expand, into the vicinity of the S&E.

Integrated circuit (IC) technology has made components smaller and cheaper but most of these components are mainly aimed at the consumer market, not high temperature applications. Therefore, proper encapsulation of S&E is a critical step in the reliability process. Figure 6.1a shows a schematic of an S&E configuration. ICs generally have plastic packaging but are prone to failures as explained before. The most common approach for replacing plastic packaging is re-packaging the S&E in ceramic or custom ceramic components [7, 9, 10, 12]. Extensive testing is however required after re-packaging due to mismatch with the original footprints, new pin layouts, and the need to validate their expected superior performance compared with components encapsulated in plastic. The die, where the S&E are fabricated on, is connected to the pins of the IC by a process known as wire bonding. The die is normally silicon (Si), which has excellent thermal conductivity, but the wires used for wire bonding, the pins and the soldering between the pins and a PCB board and the glue holding the die in the packaging are susceptible to failure. To minimize failure rates gold (Au) and aluminum (Al) are used for wire bonding, high temperature alloy materials are used for soldering, and epoxies or adhesives are used to glue the S&E inside the package [7, 9, 10, 12–14]. MCMs such as Co-fired ceramic (HTCC) and (Alumina) boards are used to combine multiple ICs into a single system level unit. They are generally plated with Al and Au for soldering and wire-bonding and the dies on these boards are processed independently and assembled into a single device as a final step. These hybrid boards are interconnected with each other in 2D or 3D layers using ceramic single inline

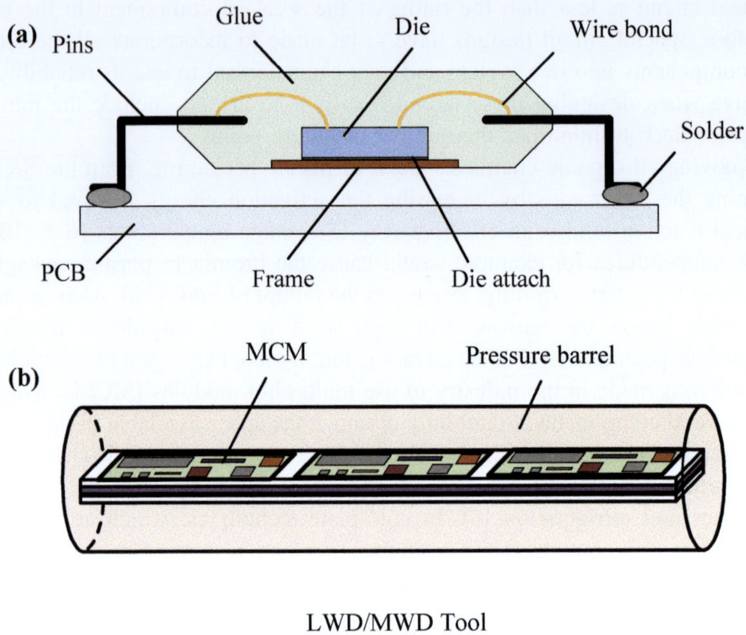

Fig. 6.1 **a** Schematic of an integrated circuit soldered onto a printed circuit board. **b** Multi chip module with sensors and electronics mounted inside a MWD/LWD tool

package headers on brazed pins (BeNi contacts). BeNi is commercially available and is a standard technology for high temperature packaging. HTCC packages have excellent mechanical rigidity, thermal dissipation and hermeticity, important features in harsh, high temperature applications. To minimize flexing MCMs a stiffening component such as a bridge over the boards or side rails is incorporated into the assembly. The MCMs are mounted inside the MWD/LWD tool by probe mounting, as shown in Fig. 6.1b, or collar sub-wall pocket mounting. In probe mounting the MCMs are inside a pressure barrel that provides minimal compression to secure the MCMs in place and, is supported by centralizers (Optimize drilling operations). In sub-wall mounting a sealing hatch is bolted over the top after the MCMs is secured in its appropriate pocket [9, 10].

Testing and reliability of fabricated and packaged S&E is the final and most critical step before the tools are sent out to drilling rigs [7, 9–11]. All S&E are repetitively characterized, and production tested, at HP/HT/HS conditions. HP/HT conditions are simulated in pressure chambers and ovens while shock for example can be simulated by free-fall dropping a module at a given height to provide quality assurance. Simulations are performed to analyze the effect vibration and resonance has on the S&E and then tested in the lab. Initially both temperature and vibration tests are done separately until failure, followed by testing both parameters together until failure. This is followed by root cause analysis and identification of the failed location and component, repairing and re-testing until reliability goals are achieved. Once evaluated for reliability, mean time between failure (MTBF) values are assigned to each S&E component and module, where the MTBF is the ratio of the number of hours a component or module can operate without failure to the number of failure hours. These MTBF values are then projected on to the LWD/MWD tools.

6.2.2 Ultra-High Temperature/Pressure Environments

Increasing the operating temperature limits of current LWD/MWD tools or using similar tools in geothermal wells will require development of a new platform for the electronics used in these tools [15, 16]. Silicon-on-insulator (SOI) is an alternative technology to Si that can be utilized for HP/HT/HS S&E [17–21]. Compared to bulk Si technology, SOI significantly reduces leakage currents and variations in device parameters, improves carrier mobility, electromigration between interconnects and dielectric breakdown strength [6, 20–22]. Silicon Carbide (SiC) based electronics is another emerging technology but has superior properties to silicon based electronics that makes it an ideal candidate for harsh environment applications, which are thermally, mechanically and chemically aggressive [23–26]. Silicon has a bandgap of 1.12 eV and loses its pn junction characteristics around 200–225 °C. SiC on the other hand has a bandgap of >3 eV and can tolerate temperatures up to 600 °C. Compared with Si, SiC also has a higher melting point (1420 °C vs. 2830 °C), thermal conductivity (1.5 W/cm K vs. 5 W/cm K), Young's modulus (190 GPa vs. 450 GPa) and a higher breakdown electric field (2.5×10^5 vs. 1×10^6). Moreover, tests done at 850 °C for 100 h showed that the SiC had an oxidizing thickness of only 50 nm compared to 300 nm for Si.

However, both SOI and SiC technologies are technically immature and is more expensive than Si technology. Moreover SOI and SiC technologies require new platforms, footprints and layouts with extensive programming, new circuit designs using SOI/SiC chips as well as suitable electronic components to maximize the advantages offered by SOI/SiC technologies. Finally, and most importantly, oil/gas industries have spent a significant amount of money for the design, testing, manufacture and maintaining reliability of the currently available tools that have 5–15 year life cycles. Therefore, spending more money for extensive modifications to the current processes and layouts of fabrication and packaging is most probably not viable for the majority of companies from an economic point of view until present tools become obsolete.

References

1. Z.A. Khan, B.K. Behera, V. Kumar, P. Sims, World Oil 75 (2012)
2. R. Beckwith, J. Petrol. Technol. **65**, 56 (2013)

3. C. Avant, S. Daungkaew, B.K. Behera, S. Danpanich, W. Laprabang, I.D. Santo, G. Heath, K. Osman, Z.A. Khan, J. Russell, Oilfield Rev. **24**, 4 (2012)
4. K. Panetta, Electronic Component Magazine (2015). Available online at https://www.ecnmag.com/blog/2015/07/what-harshest-environments-components-whats-best-way-deal-those-challenges
5. T. Baird, T. Fields, R. Drummond, D. Mathison, B. Langseth, A. Martin, L. Silipigno, Oilfield Rev. **5**, 15 (1993)
6. J.M. Li, Vectron International (2004). Available online at http://application-notes.digchip.com/073/73-41575.pdf
7. R. Beckwith, J. Petrol. Technol. **65**, 42 (2013)
8. R. Dirksen, J. Petrol. Technol. **63**, 24 (2011)
9. I. Ahmad, O. Akimov, P. Bond, P. Cairns, T. Gregg, T. Heimes, G. Russell, F. Wiese, in *The Abu Dhabi International Exhibition and Conference* (2013)
10. I. Ahmad, O. Akimov, P. Bond, P. Cairns, T. Gregg, T. Heimes, G. Russell, F. Wiese, in *Offshore Technology Conference Asia* (2014)
11. B. Parmentier, O. Vermesan, L. Beneteau, Design of high temperature electronics for well logging applications. Available online at https://www.sintef.no/globalassets/upload/ikt/9032/hiten-2003-paper-sintef-slb-july7.pdf
12. M. Watts, in *Proceedings of the International High Temperature Electronics Conference* (2008), pp. 279–286
13. K. Bybee, J. Petrol. Technol. **52**, 56 (2000)
14. Q. Tan, H. Kang, J. Xiong, L. Qin, W. Zhang, C. Li, L. Ding, X. Zhang, M. Yang, Sensors **13**, 9896 (2013)
15. D. MacGugan, in *International Conference and Exhibition on High Temperature Electronics Network* (2013)
16. J. Vedum, M.H. Røed, S. Kolberg, M. Hjelstuen, A.E. Liverud, Ø.N. Stamnes, in *Proceedings of the International Microelectronics Assembly and Packaging Society* (2017), pp. 000011–000019
17. D.R.M. Woo, J.A.K. Yun, Y. Jun, E.W.L. Ching, F.X. Che, in *Electronics Packaging Technology Conference* (IEEE, 2014), pp. 16–21
18. F. Udrea, S.Z. Ali, M. Brezeanu, V. Dumitru, O. Buiu, I. Poenaru, M.F. Chowdhury, A. De Luca, J.W. Gardner, in *International Semiconductor Conference* (2012), pp. 3–10
19. M.A. Huque, B.J. Blalock, C. Su, R. Vijayaraghavan, S.K. Islam, L.M. Tolbert, in *Proceedings of International Conference and Exhibition on High Temperature Electronics* (2008)
20. R.A. Normann, J.A. Henfling, D.J. Chavira, in *Proceedings World Geothermal Congress* (2005)
21. R.A. Normann, Drill. Contractor 63 (2007)
22. J.D. Rogers, B. Ohme, R.A. Normann, The American Oil & Gas Reporter (2005)
23. G.W. Hunter, R.S. Okojie, P.G. Neudeck, G.M. Beheim, G.E. Ponchak, G. Fralick, J. Wrbanek, L.Y. Chen, Electr. Eng. 8 (2006)
24. G.W. Hunter, P.G. Neudeck, G.M. Beheim, R.S. Okojie, L. Chen, D. Spry, A. Trunek, in *The 212th Electrochemical Society Meeting* (2007). Available online at https://ntrs.nasa.gov/search.jsp?R=20090013999
25. J. Yang, Sensors **13**, 2719 (2013)
26. J. Yang, Sensors **13**, 1884 (2013)

Chapter 7
New Technologies

Abstract This chapter presents novel ways of obtaining downhole parameters and remote activation of downhole tools utilizing uniquely designed instruments. Radio frequency identification (RFID) tags are employed for drillstring component tracking and actuating downhole valves and reamers. At/Near-bit/Short-hop sensing instruments are located in the vicinity of the drill bit or in the drill bit sub. Being directly behind the drill bit, right the bottom of the well, allows them to obtain measurements where rapid changes in geology and lithology of rock formations occur. The drilling micro-chip explores a new concept, where a mobile chip travels down a well to the bottom, and then back up again to the surface, with drilling fluid flow. During this journey the micro-chip acquires measurements resulting in the complete profile of a wellbore.

7.1 Radio-Frequency Identification

Radio frequency identification (RFID) technology has its origins in radar and radio technology used in World War II but only moved into mainstream applications in the areas of labeling manufactured goods and materials and animal tracking in the early 2000s [1]. RFID is either passive or active. A passive RFID tag does not require a power source to operate and, consists of a semiconductor chip with an antenna encapsulated in glass or a polymer to protect it from the environment. Since passive tags do not require a power source they can be used indefinitely. An active RFID tag, also known as a reader or transponder, requires a power source such a battery to operate and also consists of a semiconductor chip with electronics and an antenna encapsulated for protection. In a typical operation a passive tag is energized by the antenna of the reader when they are in the vicinity of each other. The passive tag transmits its unique identification number and specific instructions to the reader and the reader performs an operation based on the instructions. The readers have the ability to only respond to a specific identification code and to ignore other codes, and also to eliminate repetition of operations by only accepting a unique code once.

C. P. Gooneratne et al., *Instruments, Measurement Principles and Communication Technologies for Downhole Drilling Environments*, Smart Sensors, Measurement and Instrumentation 32, https://doi.org/10.1007/978-3-030-04900-3_7

The oil/gas industry first started adapting RFID technology in the late 2000s, and in drilling it is currently utilized for three specific operations. The first application is in drillpipe supply change management, where tracking drilling components is an effective barrier to preventing fatigue based failures [2]. In the second operation, as shown in Fig. 7.1a, an RFID tag programmed at the surface is dropped into the drillstring assembly and travel downhole with the drilling fluid. The drillstring assembly has a circulating sub with bypass valves and an RFID reader, located at a certain distance above the drill bit. The reader antenna energizes the RFID tag and then switches to receiving mode to verify the unique identification code and receive the instructions programmed into the tag. The instructions are then processed by the electronics in the reader to open the bypass valves, which allows circulation directly into the wellbore [3–7]. This operation is used to place the bypass valves at a selected depth below fractures and pump lost circulation material (LCM) through them to plug the fractures, as shown in Fig. 7.1a. After the operation another RFID tag can be dropped to close the valves or the previous tag can be programmed with instructions to close the valves after a certain period of time. Similar operations can be performed to change the drilling fluid or pump cement into the wellbore at

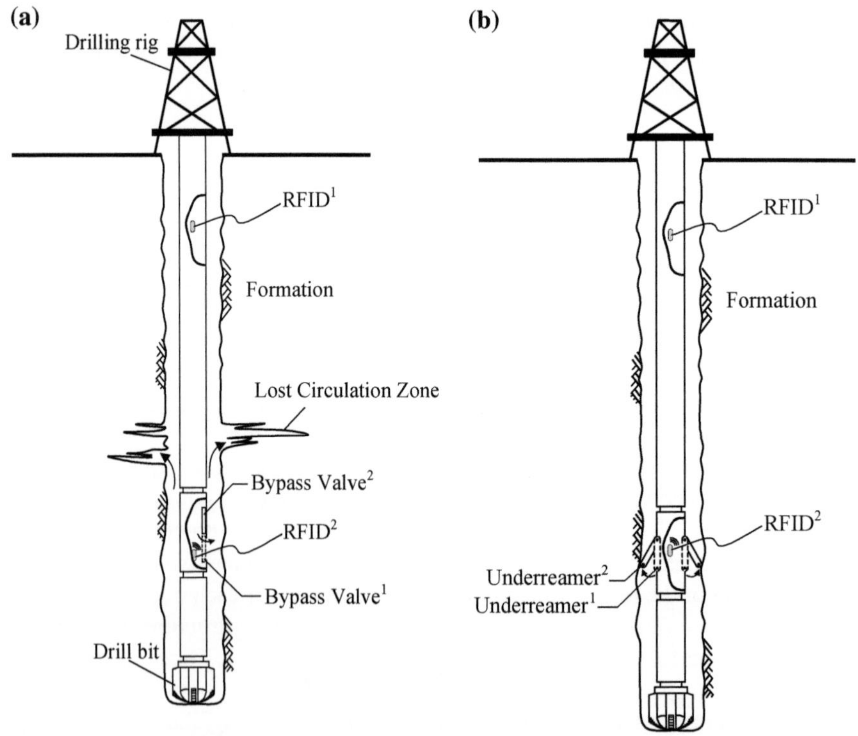

Fig. 7.1 **a** An RFID activated bypass valve opens the drillstring and allows lost circulation material to flow into and plug fractured rocks. **b** An RFID activated reamer that can expand its size to increase the size and/or smoothen a wellbore

desired depths. To minimize power consumption of the reader the energize mode of the reader can be activated by pressure cycles from the surface after or right before the RFID tags are pumped down the drillstring assembly. Pressure cycles applied from the surface can also be used to activate the reader as a contingency method in the event a circulating flow is not possible.

The third application is activating a drilling underreamer, as shown in Fig. 7.1b. A drilling underreamer is a tool with cutters that is located behind a drill bit. It is utilized to enlarge, smoothen and condition a wellbore for running casing or completion equipment without any restrictions. Rather than pulling the drillstring assembly out of the well when problems arise downhole, a reamer that is activated by an RFID is run inside the well with the drillstring assembly. An RFID tag dropped from the surface travels with the drilling fluid towards the drilling underreamer. Once in the locality of the underreamer the RFID tag is energized by the underreamer antenna and the RFID tag in return transmits its unique identification code and specific instructions to the underreamer. The underreamer then extends and drills through with the drill bit. Another RFID tag can be dropped from the surface to retract the underreamer. RFID tags can be programmed to extend or retract reamers in several finite steps depending on the desired diameter of the wellbore [3, 4, 6, 8, 9]. Also, the underreamer can be activated multiple times, at selective zones, or multiple underreamers can be activated at multiple and different times and zones.

The procedure normally used for activating bypass valves involves dropping an activation ball to open the two side ports, pumping the LCM/Cement/Fluid and then dropping two deactivation balls to close the ports. Applying pressure from the surface releases all three balls that fall into a ball catcher cage below. A similar procedure is used for triggering a reamer. The RFID activation method has several advantages over ball activation method. Remote activation and no restrictions inside the drillstring resulting in a larger flow area for the drilling fluids, allows any logging instrument to pass through the drillstring without restriction, there is no risk of LCM damaging the MWD unit or RSS tools below the bypass valve, can be used to perform multiple operations at one depth or several depths with bypass valves placed at one location or multiple locations on the drillstring, avoids extra trips to the surface to remove the balls or reamer from the drillstring assembly.

Areas for further development of RFID technology are the limits imposed by the battery on the operational temperature and time of the reader, compatibility with different tools, a vertical orientation of $\pm 45°$ required while passing through the reader antenna, activation methods and timings, evaluation of contingency options and the need for specific configurations of, and modifications to, existing tools, or design of new tools, in the drillstring assembly to accommodate RFID technology.

7.2 At/Near-Bit/Short-Hop

At/Near-bit sensing technology allows drillers to obtain measurements from 0.3 to 3 m behind the drill bit rather than 15 m for preceding MWD/LWD tools. Placement of sensors much closer to the drill bit results in significant reduction in the actual downhole measurement lag experienced by placing the sensors further above the drill bit. Obtaining real-time measurements much closer to the downhole environment, which gives a better representation of the position of the bit, means rapid changes in geology and lithology of rock formations can be met with quick responses, well navigation can be performed faster and the well can be steered in the desired direction more accurately. Moreover, obtaining drilling dynamics data near the bit can significantly improve the drilling efficiency. The advantages of at/near bit sensors are more pronounced in complex geologic settings. Figure 7.2 shows a schematic of an at/near-bit tool.

Short hop units are used when the location of a hydraulic mud motor prevents sensors from below the motor being easily connected or wired to the measurement-while-drilling (MWD) system located above the motor due to the thin housing of the power section [10]. Short hop units are applied most commonly for a rotary drilling assembly and used to 'hop' between the MWD above the hydraulic motor and the instrumented drill bit. However, they can also be used in motor driven

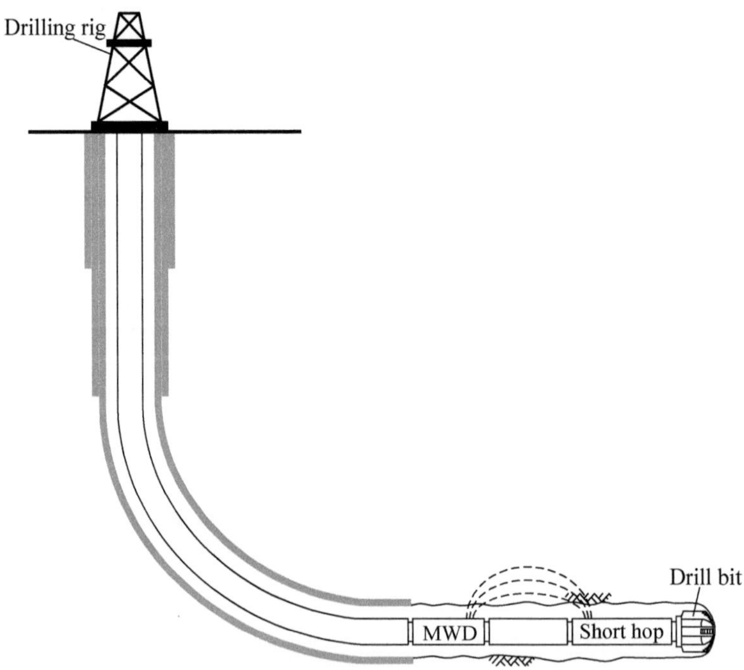

Fig. 7.2 Sensors near the bit obtain measurements very close to the bottom of the hole and transfer this data by electromagnetic wireless data telemetry

rotary steerable system (RSS) applications. The short hop unit and MWD unit are two different subs. An electromagnetic wireless telemetry unit in the short hop unit transfers the measurement data, or the measurement data 'hops', to the MWD unit after which the data is transferred to the surface, normally by mud pulse (MP) telemetry in the MWD unit.

The early at/near-bit tools in the 90s measured azimuthal resistivity, gamma ray, inclination and toolface and revolutions per minute (RPM) [11]. Gamma ray and resistivity provides early indication of changes in geology and lithology of rock formations, inclination and toolface provides more accurate well trajectory, and RPM shows the actual revolutions in the vicinity of the drill bit rather than the RPM at the surface, which is always higher than the true value. The drill bit acts a measuring electrode and a set of electrodes are connected directly to the bit for at/near-bit resistivity measurements and sensors can be placed near, middle or far depending on shallow, medium, deep resistivity measurements. Current gamma ray instruments are able to obtain azimuthally sensitive at-bit images for steering a well, especially in thin reservoirs to maximize access and exposure to hydrocarbons, and resistivity tools are able to look ahead of the bit and determine changes in resistivity of the formation, while drilling [12–15].

Until the mid-2000s drilling dynamics instruments measured real-time data but were confined to locations well above the drill bit. The first at-bit drilling dynamics instruments located a few meters above the drill bit measured only temperature and acceleration at the bit. Current instruments are compact enough to be slotted inside the drill bit sub, as shown in Fig. 7.3, and are able to measure torsional, axial and lateral vibrations, WOB, TOB, RPM, temperature, and inner and outer drillstring pressure [16–18]. These instruments are only able to record and store the measurements in memory and can be retrieved from the drill bit sub when the drillstring assembly is pulled out of the wellbore. The recorded data provides valuable

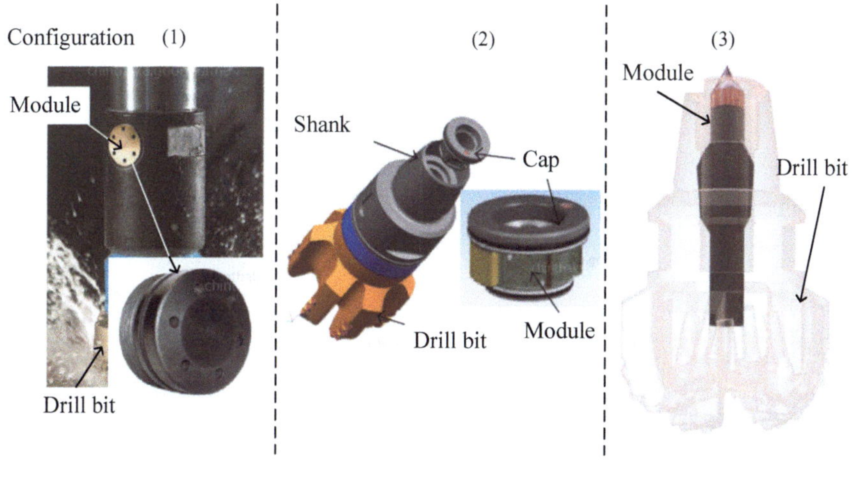

Fig. 7.3 At/near-bit sensor modules in three different configurations above the drill bit [16]

information on the entire drilling system, driving both short and long term improvements in drilling methods and efficiency. This data then can be used to create drilling roadmaps for other wells on the same field or to predict behavior in similar fields for improving drilling performance.

While this at/near bit technology obtains measurements that better represent the downhole environment, the number of instruments that can be placed near the drill bit is limited and the instruments being exposed to the extremely harsh conditions at the influences their performance and lifespan compared to instruments located further above the drill bit. If the drilling assembly contains a hydraulic motor above the at/near bit instruments, additional data transmission means (data hopping from instruments to telemetry unit) are needed to ensure real-time data transmission to surface, which adds considerable technical complicities and therefore prone to problems while operating the tools.

7.3 Drilling Micro-Chip

The drilling micro-chip is a smart, miniature mobile device with integrated sensors that can be injected into wells and recovered at the surface to obtain near real-time temperature and pressure measurements in wells. The micro-chip consists of a battery, temperature/pressure sensors, a microcontroller, memory and a data transmission module encapsulated in spherical ball to protect the components from the harsh downhole environment. Advancements in MEMS (micro-electro-mechanical systems) technology have paved the way for building these devices at the mm/micro/nano scale and then integrating them together to create low-cost, miniature, smart, micro-chips [19–27]. Micro-chips are useful in downhole applications because they can be deployed down a well with drilling fluid flow to flow right to the bottom of a well and up again to the surface therefore, providing a full profile of the wellbore. Table 7.1 shows the advantages of the micro-chip compared

Table 7.1 Comparison of existing downhole hole data acquisition methods [30]

Wireline logging	MWD/LWD	Intelligent/wired drill pipe	Drilling microchip
Since 1927	Since 1978	Since 2005	Since 2011
Moderate cost	High cost	High cost	Low cost
Whole profile logging	Bottom hole logging	Whole profile logging	Whole profile logging
High data speeds	0.5–10 bits/s	Megabits/s	Saved on memory
After-the-fact	Real-time	Real-time	Near real-time with a delay
Powered through cable	Batteries and mud turbine	Powered through drill pipe	Battery
Multi-sensor	Multi-sensor	Multi-sensor	Temperature, pressure

with WL logging, MWD/LWD and wired telemetry. The micro-chip does not provide real-time data like MWD/LWD or wired telemetry but provides near real-time data while drilling, which is an advantage compared with WL logging. It is also a much cheaper and easier method than wired telemetry to obtain measurement data for the whole wellbore since no modifications are required to the standard drilling equipment. The micro-chips are also lightweight and can be mass fabricated at low cost.

The 1st generation micro-chip was introduced in 2011 and a modified 2nd generation micro-chip was released in 2015 [28, 29]. The operation procedure for both micro-chips is the same. As shown in Fig. 7.4a, once dropped directly into the drillstring assembly the micro-chips flow down towards the drill bit, where it then exits through the drill bit nozzles and goes up the annulus all the way to surface. During this process measurements are recorded sequentially at preprogrammed time intervals and saved in the memory. The micro-chips are recovered from the drilling

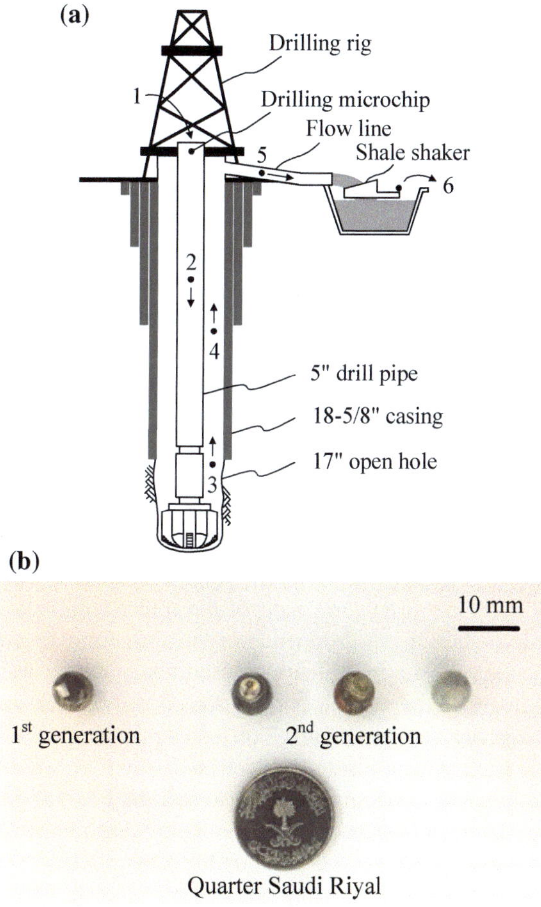

Fig. 7.4 **a** Deployment of the micro-chip from the surface, into the drillstring assembly, out the annulus and back onto the surface to measure temperature of the whole well profile. **b** First and second generation micro-chips, where the latter has one sensor for either temperature or pressure and the former had both sensors in one chip [30]

fluid flow at the surface, near the shale shaker, which is the area where the rock formation cuttings from the well are filtered. The data stored in the memory of the micro-chip is then downloaded and the well depth information of each measurement point is calibrated by the timestamp of the recorded data and the drilling fluid flow rate. The fabricated 1st and 2nd generation micro-chips are shown in Fig. 7.4b. The spherical shell is made of high strength material with physical/chemical stability and the density of the shell was used to adjust the specific gravity of the micro-chip so that it could easily circulate with the drilling fluids. The 1st generation micro-chip was able to communicate with an external transceiver wirelessly for recharging, initialization, and downloading data. It had fiber-optic sensors for temperature and pressure measurement, analog-to-digital converter, an 8-bit microcontroller and a 10 kb flash memory on a system-on-chip package. The field tests performed with the 1st generation micro-chips were the first time a mobile instrument was able to measure temperature and pressure inside a well. The main limitations of the 1st generation micro-chip were the high power consumption of the fiber-optic sensors and the in-built transceiver for wireless communication. This along with the high temperature downhole resulted in the micro-chip only surviving for 40 min inside the well. Also, coupling of the optical media resulted in interference that affected the accuracy of measurements.

In addressing the above limitations the 2nd generation micro-chip was fabricated with a wired connection and only one sensor, either temperature or pressure, to prevent two sensors from interfering with each other. The battery capacity was also doubled from 1 to 2 mAh, the power consumption was reduced from 500 to 200 μW and the battery life was increased from 2 to 10 h. Figure 7.5 shows the significant improvement in the accuracy of the measurements when only one sensor was used in the micro-chip. The error in temperature measurements decreased from ±8 to ±1.8 °F and the error in pressure measurements decreased from ±1000 to ±60 psi.

Preliminary field tests performed in 2013 demonstrated the increased accuracy of the 2nd generation micro-chips but tests performed in 2016 proved that the micro-chip was capable of obtaining measurements over the whole profile of a wellbore, 2350 ft deep [30]. Figure 7.6a shows the field results of the 2nd generation micro-chip. 20 micro-chips were pumped down the drillstring assembly while drilling and 6 were recovered at the surface, with 3 intact and 3 broken. Data downloaded from the 2 micro-chips are identical as shown in Fig. 7.6a. Sections 1–6 show the circulation path of the micro-chip, where section 1 indicate when the micro-chips were dropped into the drillstring assembly, section 2 when they are flowing inside the drillstring assembly, sections 3 and 4 represent the micro-chips flowing up the annulus and sections 5 and 6 when the micro-chips were retrieved at the shale shaker. Figure 7.6b shows a detailed view of sections 1–4 when the micro-chip was inside the well and it can be seen that the temperature increases from when it was first dropped in section 1 to a maximum temperature at the end of section 2, at the bottom of the hole. The temperature drops in section 3 as the micro-chips go up the annulus but a continuous decrease was not observed most probably due to the heat conduction of the steel casing. The flow rate and area are

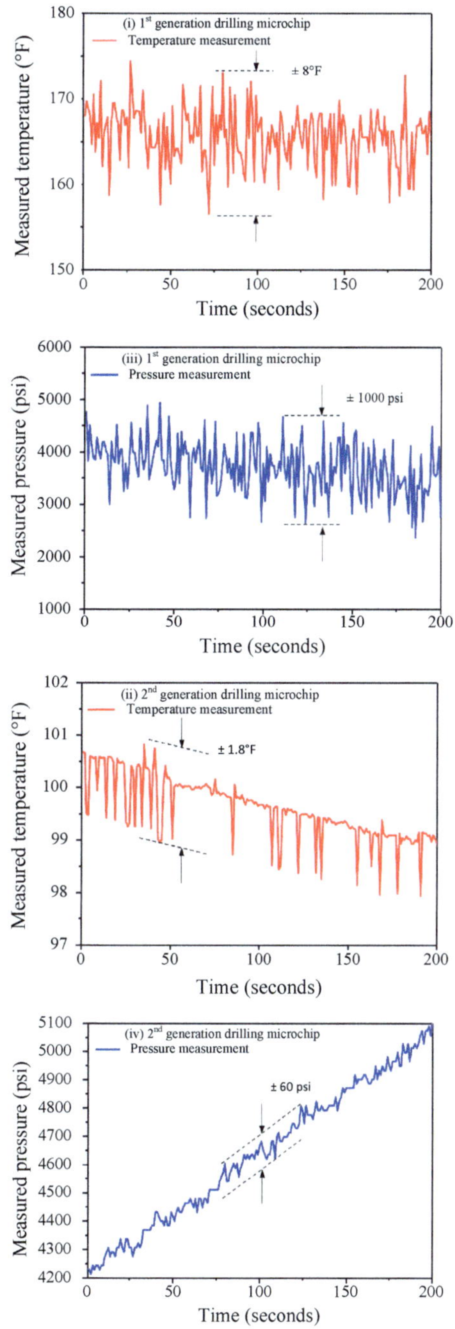

Fig. 7.5 Experimental results showing a significant improvement in the noise level in the second generation micro-chips (two graphs at the bottom) compared with the first generation micro-chips (two graphs at the top) [30]

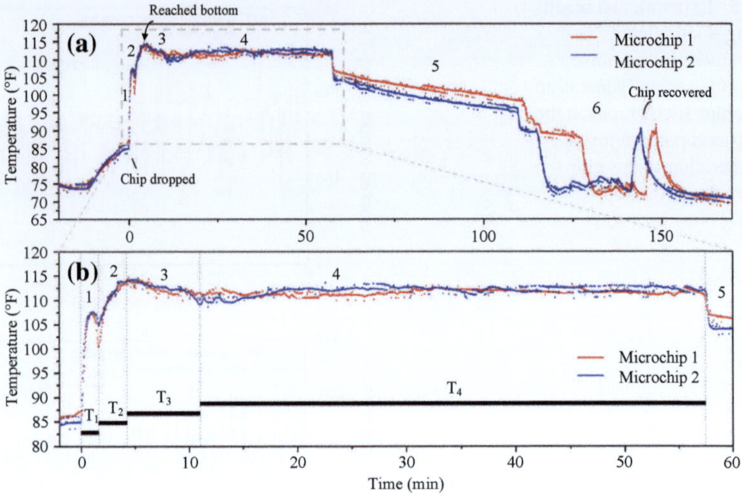

Fig. 7.6 **a** Temperature field test result of the 2nd generation drilling microchip. **b** A detailed view of the temperature measurement from sections 1–4 [30]

used to calculate T2–4 for flow sections 2–4. While T2 and T3 agreed with the theoretical times T4 was longer, which was most probably due to rock formation cuttings reducing the flow speed of the micro-chips when they were flowing up the annulus. Further work is now being performed to improve the recovery method by alternative methods to the current visual retrieval method and making the micro-chips smaller so that they can pass through a wide variety of drill bit nozzles.

References

1. R. Want, IEEE Pervas. Comput. **5**, 25 (2006)
2. M. Grinrod, M. Vonlanten, A.T. Haaland, R. Gaaso, in *SPE/IADC Drilling Conference* (2013)
3. P.M. Snider, T. Doig, in *SPE/ICoTA Coiled Tubing and Well Intervention Conference and Exhibition* (2008)
4. E. Murdoch, E. Valverde, R. Sharma, C. Wreden, A. Goodwin, J. Osei-Kuffour, K. Kimmitt, C. Loonstra, B. Ghaempanah, D. Reagins, in *SPE Intelligent Energy International Conference and Exhibition* (2016)
5. E. Valverde, R. Reitsma, T. Hutchings, in *SPE Annual Technical Conference and Exhibition* (2015)
6. E. Valverde, E. Murdoch, C. Wreden, A. Goodwin, J. Osei-Kuffour, K. Kimmitt, C. Loonstra, B. Ghaempanah, D. Reagins, in *Offshore Technology Conference* (2017)
7. A. Goodwin, R. Reitsma, J. Osei-Kuffour, in *SPE Annual Technical Conference and Exhibition* (2016)
8. T. Laird, L. Gonzalez, E. Valverde, A.C. Odell, in *SPE Annual Technical Conference and Exhibition* (2011)

9. A. Goodwin, C. Salminen, E. Valverde, in *SPE Annual Technical Conference and Exhibition* (2015)
10. A. Wright, J. Snyder, in *SPE/IADC Drilling Conference and Exhibition* (2014)
11. S. Bonner, T. Burgess, B. Clark, D. Decker, J. Orban, B. Prevedel, M. Lüling, J. White, Oilfield Rev. **5**, 44 (1993)
12. J.L. Pitcher, D.B. Schafer, P. Botterell, in *SPE/IADC Drilling Conference and Exhibition* (2009)
13. A.J. Wheeler, T. Billings, A. Rennie, R. Lee, R. Little, C. Huiszoon, P. Boonen, in *SPWLA 53rd Annual Logging Symposium* (2012)
14. R. Flook, W. Alexander, D. List, B. Sencenbaugh, B. Enoch, A.J. Wheeler, A. Starkey, A. Rennie, J. Petrol. Technol. **65**, 30 (2013)
15. M.V. Constable, F. Antonsen, S.O. Stalheim, P.A. Olsen, Ø.Z. Fjell, N. Dray, S. Eikenes, H. Aarflot, K. Haldorsen, G. Digranes, J. Seydoux, in *SPWLA 57th Annual Logging Symposium* (2016)
16. G. Dong, P. Chen, Shock and Vibration (2016). Available online at http://dx.doi.org/10.1155/2016/7418635
17. L.W. Ledgerwood III, J.R. Jain, O.J. Hoffmann, R.W. Spencer, SPE Drill. Complet. **28**, 254 (2013)
18. L.A. Lines, C.L. Mauldin, J.W. Hill, R.A. Aiello, in *SPE Annual Technical Conference and Exhibition* (2014)
19. O. Yassine, C.P. Gooneratne, D. Abu-Smara, F. Li, H. Mohammed, J. Merzaban, J. Kosel, Biomicrofluidics **8**, 034114 (2014)
20. A.I. Zia, A.R. Mohd Syaifudin, S.C. Mukhopadhyay, I.H. Al-Bahadly, P.L. Yu, C. P. Gooneratne, J. Kosel, T.S. Liao, in *Instrumentation and Measurement Technology Conference* (2013)
21. C.P. Gooneratne, I. Giouroudi, J. Kosel, Sensor Lett. **10**, 770 (2012)
22. J.W. Judy, Smart Mater. Struct. **10**, 1115 (2001)
23. P. French, G. Krijnen, F. Roozeboom, Microsyst. Nanoeng. **2**, 16048 (2016)
24. C.P. Gooneratne, O. Yassine, I. Giouroudi, J. Kosel, IEEE Trans. Magn. **49**, 3418 (2013)
25. C.P. Gooneratne, C. Liang, J. Kosel, Microelectron. Eng. **88**, 1757 (2011)
26. C.P. Gooneratne, A. Kurnicki, S. Yamada, S.C. Mukhopadhyay, J. Kosel, PLoS ONE **8**, e81227 (2013)
27. Y. Huang, A.S.S. Vasan, R. Doraiswami, M. Osterman, M. Pecht, IEEE Trans. Device Mater. Reliab. **12**, 482 (2012)
28. M. Yu, S. He, Y. Chen, N. Takach, P. LoPresti, S. Zhou, N.M. Al-Khanferi, in *SPE Annual Technical Conference and Exhibition* (2012)
29. Z. Shi, Y. Chen, M. Yu, S. Zhou, N. Al-Khanferi, in *SPE Digital Energy Conference and Exhibition* (2015)
30. B. Li, C.P. Gooneratne, M.S. Badran, Z. Shi, in *SPE/IATMI Asia Pacific Oil & Gas Conference and Exhibition* (2017)

Chapter 8
Future Outlook and Perspectives

Abstract This chapter discusses the current technological status of the drilling industry and the direction in which it is moving in to be in alignment with the fourth industrial revolution (4IR). To adapt and keep up with the resulting shifts and disruptions, and fully reap benefits from the technologies, associated with 4IR, it is believed that drilling will have to take a technology leap rather than a technology step.

8.1 History of Technology Evolution in Drilling

Drilling is all about reaching a hydrocarbon reservoir thousands of feet below the ground as fast and safely as possible by managing pressures. Since the drilling operation cannot be seen engineers have designed a unique set of instruments over the years to understand and 'get a feel' for the rock formations and the downhole environment they are drilling through, along with navigation instruments to guide them in any direction to reach hydrocarbon reservoirs. Measurement instruments have evolved from the first electrical log 90 years ago to instruments that now provide real-time measurements of a large number of parameters in a downhole environment. Until the late 1970s and early 1980s drilling had to be ceased to obtain measurements downhole with WL logging. This was a time-consuming, expensive procedure, which also ran the risk of getting the WL tool stuck inside the well and affecting the stability of the well. A technology leap was made when a method called mud pulse telemetry was successfully invented to transmit measurement data while drilling and this enabled the driller to obtain real-time measurement data during the drilling operation. Over the years, especially when hydrocarbon reservoirs were relatively easy to access, these instruments have played a vital role by providing measurements data on geology and lithology of different rock formations, characteristics of rock formations such as porosity, pore geometry, permeability, resistivity, saturation and density and directional data such as azimuth and inclination, to optimize drilling efficiencies and maximize hydrocarbon production.

C. P. Gooneratne et al., *Instruments, Measurement Principles and Communication Technologies for Downhole Drilling Environments*, Smart Sensors, Measurement and Instrumentation 32, https://doi.org/10.1007/978-3-030-04900-3_8

3. C.G. Christodoulou, Y. Tawk, S.A. Lane, S.R. Erwin, Proc. IEEE 100, 2250 (2012)
4. C.P. Gooneratne, C. Liang, J. Kosel, Microelectron. Eng. 88, 1757 (2011)
5. C.P. Gooneratne, I. Giouroudi, J. Kosel, Sensor Lett. 10, 770 (2012)
6. O. Yassine, C.P. Gooneratne, D. Abu-Smara, F. Li, H. Mohammed, J. Merzaban, J. Kosel, Biomicrofluidics 8, 034114 (2014)
7. M. Diaz-Michelena, Sensors 9, 2271 (2009)
8. Baker Hughes, a GE Company Investor Update. Available online at https://www.ge.com/sites/default/files/ge_webcast_presentation_11082016_0.pdf
9. M. Scott, IoT and the future of the energy industry. Available online at https://www.enlogy.com/en/technology_en/internet-of-things-energy-industry/
10. M. Verdu, The IoT will enable the gas and oil industry to cut production costs by up to $30 per barrel. Available online at http://www.iotsworldcongress.com/promise-iot-oil-gas-industry-cut-costs-30-per-barrel/

MIX
Papier aus verantwortungsvollen Quellen
Paper from responsible sources
FSC® C105338

If you have any concerns about our products,
you can contact us on
ProductSafety@springernature.com

In case Publisher is established outside the EU,
the EU authorized representative is:
Springer Nature Customer Service Center GmbH
Europaplatz 3, 69115 Heidelberg, Germany

Printed by Libri Plureos GmbH
in Hamburg, Germany